SpringerBriefs in Mathematics

SpringerBriefs in Mathematics showcases expositions in all areas of mathematics and applied mathematics. Manuscripts presenting new results or a single new result in a classical field, new field, or an emerging topic, applications, or bridges between new results and already published works, are encouraged. The series is intended for mathematicians and applied mathematicians.

For further volumes:
http://www.springer.com/series/10030

Matteo Colangeli

From Kinetic Models to Hydrodynamics

Some Novel Results

 Springer

Matteo Colangeli
Department of Mathematics
Politecnico di Torino
Turin
Italy

ISSN 2191-8198   ISSN 2191-8201   (electronic)
ISBN 978-1-4614-6305-4   ISBN 978-1-4614-6306-1   (eBook)
DOI 10.1007/978-1-4614-6306-1
Springer New York Heidelberg Dordrecht London

Library of Congress Control Number: 2013931839

Mathematics Subject Classification (2010): 76P05, 82C40, 82C70

Printed on acid-free paper

Springer is part of Springer Science+Business Media (www.springer.com)

Cum luminem sequimur,
soror itineris tenebra

Preface

The Boltzmann equation is a milestone of classical and quantum kinetic theory of gases [1–6]. It describes the evolution of a dilute gas, initially prepared in a nonequilibrium state, by means of a statistical approach which allows to disregard the detailed knowledge of the motion of the single particles. In the absence of external forces, the gas undergoes a process of relaxation to equilibrium, whose mathematical essence is captured by the celebrated H-Theorem, which also marked the first clear onset of irreversibility in classical particle systems. On the other hand, a different perspective in the behaviour of a fluid is offered by the laws of continuum physics, which found a self-contained settlement in the equations of hydrodynamics. This work is concerned with the ambitious and long-standing task of linking the two different levels of description, the kinetic and the hydrodynamic. The question arises as to whether this is actually an original ambition. The Grad's moment method, for instance, which still spreads its influence on the modern theory of Extended Irreversible Thermodynamics [7], dates back to the late 1940s of the last century. Earlier attempts are represented by Hilbert's procedure and by the Chapman-Enskog expansion, which constitute an important success in kinetic theory, as they allowed to derive the hydrodynamic laws from the Boltzmann equation and provided consistent expressions for the transport coefficients. However, almost a century of effort to extend the hydrodynamic description beyond the Navier–Stokes–Fourier approximation failed even in the case of small deviations around the equilibrium, due to the onset of instabilities which prevent the use of the hydrodynamic solutions [8, 9]. A different route, in kinetic theory, is represented by the recent Invariant Manifold method [10]. This technique, based on the computation of a slow invariant manifold in the space of distribution functions, provides hydrodynamic equations which are stable and remain valid also at short length scales, provided that the condition of local equilibrium holds. Thus, the purpose of this work is to offer a short survey over a field of active research, which aims at bridging time and length scales, from the particle-like description inherent in the Boltzmann theory up to the hydrodynamic setting. Our plan is to perform a bottom-up approach, which steps from the statistical foundations of the Boltzmann equation to the spectral properties of hydrodynamic fluctuations. Our natural inclination is to shape things as they were rooted and naturally emerging each from the other: i.e., by showing, for instance, how typical kinetic equilibration rates affect the normal modes of hydrodynamic fluctuations and, also, how the hydro-dynamic setting may be successfully extended to length scales comparable with the mean free path. A great scientist of our times, J. L. Lebowitz, expressed, in [11], his own surprise in considering that, in spite of the hierarchical structure of Nature, characterised by a variety of time and length scales, it is still possible, to some extent, to discuss the various levels of description independently of one another. *"Thus, arrows of explanations between different levels always point from smaller to larger scales, although the origin of higher level phenomena in the more fundamental lower level laws is often very far from transparent"*. Our overall impression, which we would like to share with the reader, is that a unifying approach is finally starting to take shape.

References

1. S. Chapman and T. G. Cowling, *The Mathematical Theory of Nonuniform Gases*, (Cambridge University Press, New York, 1970)
2. C. Cercignani, R. Illner and M. Pulvirenti, *The Mathematical Theory of Dilute Gases* (Springer, Berlin, 1994)
3. C. Cercignani, *Theory and Application of the Boltzmann Equation* (Scottish Academic Press, Edinburgh, 1975)
4. D. Benedetto, F. Castella, R. Esposito and M. Pulvirenti, A short review on the derivation of the nonlinear quantum Boltzmann equations, *Commun. Math. Sci.* **5**, 55 (2007)
5. D. Benedetto, F. Castella, R. Esposito and M. Pulvirenti, From the N-body Schrödinger equation to the quantum Boltzmann equation: a term-by-term convergence result in the weak coupling regime, *Comm. Math. Phys.* **277**, 1 (2008)
6. N. Bellomo Ed., *Lecture Notes on the Mathematical Theory of the Boltzmann Equation* (World Scientific, 1995)
7. D. Jou, J. Casas-Vázquez and G. Lebon, *Extended Irreversible Thermodynamics* (Springer, 2010)
8. M. Colangeli, I. V. Karlin and M. Kröger, From hyperbolic regularization to exact hydrodynamics for linearized Grad's equations, *Phys. Rev. E* **75**, 051204 (2007)
9. M. Colangeli, I. V. Karlin and M. Kröger, Hyperbolicity of exact hydrodynamics for three-dimensional linearized Grad's equations, *Phys. Rev. E* **76**, 022201 (2007)
10. A. N. Gorban and I. V. Karlin, *Invariant Manifolds for Physical and Chemical Kinetics*, Lect. Notes Phys. **660** (Springer, Berlin, 2005)
11. J. L. Lebowitz, Statistical mechanics: A selective review of two central issues, *Rev. Mod. Phys.* **71**, S346–S357 (1999)

Acknowledgments

I gratefully thank Prof. N. Bellomo for his support and my advisor, Prof. L. Rondoni, for contributing to shape my knowledge in this field through a continuous exchange of ideas, over the years.

I also acknowledge inspiring discussions with Prof. M. Pulvirenti and with Prof. E. Trizac.

A special mention, then, is for my family and for Alessandra, for their presence.

This work is dedicated to the memory of my beloved grandma, Gemma, who will be with me till the end.

Contents

Chapter 1
Introduction

A certain number of techniques have been designed in the kinetic theory of gases to derive macroscopic time evolution equations from the Boltzmann equation. Most of these methods require the single-particle distribution function to be parameterized by a set of distinguished fields, such as the hydrodynamic ones: the number (or mass) density, momentum vector, and temperature. This is a plausible assumption as long as the microscopic dynamics enjoys a vast separation of time scales and local thermodynamic equilibrium exists. Moreover, the derivation of hydrodynamics from kinetic theory is often concerned with the *hydrodynamic limit* of the Boltzmann equation. Loosely speaking, one is interested, typically, in the scaling of the Boltzmann equation with respect to some reference macroscopic length and time scales, which are expected to largely dominate the intrinsic kinetic scales. Nonetheless, it makes sense to consider the extension of the hydrodynamic description beyond the standard domain, considering reference scales comparable with the kinetic ones. This is the subject dealt with by *generalized hydrodynamics* [1, 2]. There are several delicate aspects hindering this line of investigation. A first, natural, objection points to the fact that below a certain length scale, the notion itself of "local equilibrium," which is brought about by a sufficiently large number of particle collisions, is questionable. Moreover, from the technical side, one typically deals, in this context, with perturbative methods, such as Hilbert's procedure or the Chapman–Enskog (CE) technique, which, at a certain order of truncation, may give rise to artificial instabilities [3, 4]. In particular, the CE method introduces an expansion of the distribution function in terms of a parameter, the Knudsen number, defined as the ratio of the mean free path to a representative macroscopic length. For small values of the Knudsen number, the CE method recovers the standard Navier–Stokes–Fourier (NSF) equations of hydrodynamics. In more refined approximations, referred to as the Burnett and super-Burnett hydrodynamics, the hydrodynamic modes become polynomials of higher order in the wave vector. In such an extension, the resulting hydrodynamic equations may become unstable and violate the H-theorem, as first shown by Bobylev [3] for a particular case of Maxwell molecules. This indicates that the CE theory cannot be immediately trusted away from the hydrodynamic limit. Thus, while

M. Colangeli, *From Kinetic Models to Hydrodynamics*, SpringerBriefs in Mathematics,
DOI: 10.1007/978-1-4614-6306-1_1, © Matteo Colangeli 2013

the mathematical framework concerning the hydrodynamic limit of the Boltzmann equation is well established [5, 6], there is no consolidated counterpart addressing the short-wavelength domain. On the other hand, recent technological trends in the emerging field of microfluidics [7, 8] demand an extension of hydrodynamics to the short scales and enable the development of novel methods. In this work, we discuss the invariant manifold theory, which makes it possible to derive the equations of *exact* linear hydrodynamics from the Boltzmann equation. The adjective "exact" stems from the fact that the method leads to an exact summation of all the terms occurring in the CE expansion. We will show, through the analysis of some solvable model, that the divergences of the hydrodynamic modes are actually removed by additionally taking into account the very remote terms of the expansion. This is made possible by solving a closed integral equation, called an invariance equation, that connects the microscopic evolution of the distribution function with its dynamics triggered by the set of hydrodynamic fields.

This work is structured as follows.

In Chap. 2, we will review the basic mathematical framework concerning the statistical description of a classical many-particle system. We will also discuss a heuristic derivation of the Boltzmann equation and some of its scaling forms relevant to our approach.

In Chap. 3, we will outline some standard model reduction techniques, which allow us to derive hydrodynamics from the Boltzmann equation.

In Chap. 4, we will introduce the correlation function formalism and discuss the role of fluctuations in fluid systems from a macroscopic standpoint.

In Chap. 5, we will derive the equations of linear hydrodynamics from certain kinetic models of the Boltzmann equation, thus providing the desired bridge between the kinetic and the hydrodynamic descriptions, valid also at short length scales.

In Chap. 6, we will provide the explicit computation of the invariant manifold for the Grad's 13-moment system and illustrate the onset of a critical length scale below which the hydrodynamic description fails.

Finally, conclusions are drawn in Chap. 7.

References

1. B.J. Alder and W.E. Alley, Generalized Hydrodynamics, *Phys. Today* **37**, 56 (1984).
2. J.P. Boon and S. Yip, *Molecular Hydrodynamics* (Dover, 1991).
3. A.V. Bobylev, Sov. Phys. Dokl. **27**, 29 (1982).
4. F.J. Uribe, R.M. Velasco, and L.S. Garcìa-Colìn, Bobylev's Instability, *Phys. Rev. E* **62**, 5835 (2000).
5. C. Cercignani, R. Illner, and M. Pulvirenti, *The Mathematical Theory of Dilute Gases* (Springer-Verlag, Berlin, 1994).
6. C. Cercignani, *Theory and Application of the Boltzmann Equation* (Scottish Academic Press, Edinburgh, 1975).
7. A. Beskok and G.E. Karniadakis, *Microflows: Fundamentals and Simulation* (Springer, Berlin, 2001).
8. D. M. Karabacak, V. Yakhot, and K. L. Ekinci, High-Frequency Nanofluidics: An Experimental Study Using Nanomechanical Resonators, Phys. Rev. Lett. **98**, 254505 (2007).

Chapter 2
From the Phase Space to the Boltzmann Equation

In this chapter, we will outline the general mathematical framework concerning the statistical description of a many-particle system in the phase space. We will then restrict our analysis to the single-particle space and will sketch the derivation of the Boltzmann equation, which was originally derived by Ludwig Boltzmann (1872) for a hard-spheres gas by merging mechanical concepts and statistical considerations.

2.1 The Phase Space Description

Let us consider a system of N identical particles, described by the coordinates $(\mathbf{q}_i, \mathbf{p}_i)$, with $i = 1, \ldots, N$, which denote, respectively, the positions and the momenta of the particles. For convenience, we will use the shorthand notation $(\mathbf{Q}, \mathbf{P})$ to refer to the entire set of coordinates in a compact phase space $\mathcal{U} \subset \mathbb{R}^{6N}$. The time evolution of this system is given by a flow $S^t : \mathcal{U} \to \mathcal{U}$, with S^t denoting a one-parameter group of diffeomorphisms. We focus hereinafter on conservative dynamical systems whose dynamics is dictated by the Hamiltonian $H(\mathbf{Q}, \mathbf{P})$, with an interaction term defined by a (smooth) pairwise potential $\Phi : \mathbb{R}^3 \to \mathbb{R}$, depending only on the distance $|\mathbf{q}_i - \mathbf{q}_j|$ between the ith and jth particles. The dynamical systems under consideration are, moreover, equipped with a probability measure $\mu(\mathrm{d}\mathbf{Q}\mathrm{d}\mathbf{P})$, with support on $\mathcal{U}$, absolutely continuous with respect to the Lebesgue measure. Hence, we suppose that a phase space density $F(\mathbf{Q}, \mathbf{P})$, positive definite and normalized, exists and is such that $F(\mathbf{Q}, \mathbf{P})\mathrm{d}\mathbf{Q}\mathrm{d}\mathbf{P}$ represents the probability of finding the system in the phase space element centered at the point $(\mathbf{Q}, \mathbf{P}) \in \mathcal{U}$. The trajectory of $(\mathbf{Q}, \mathbf{P})$ is determined by solving Hamilton's equations

$$\dot{\mathbf{q}}_i(t) = \frac{\partial H}{\partial \mathbf{p}_i}, \quad \dot{\mathbf{p}}_i(t) = -\frac{\partial H}{\partial \mathbf{q}_i}, \tag{2.1}$$

for each $i = 1, \ldots, N$.

M. Colangeli, *From Kinetic Models to Hydrodynamics*, SpringerBriefs in Mathematics, DOI: 10.1007/978-1-4614-6306-1_2, © Matteo Colangeli 2013

Let us introduce, then, the phase space functions, also called microscopic dynamical functions, $a(\mathbf{Q}, \mathbf{P}) : \mathscr{U} \to \mathbb{R}$, which, under the evolution in the phase space, are transformed, at time t, into $a(\mathbf{Q}, \mathbf{P}, t)$. It is useful to let the phase functions depend also on a vector-valued parameter $\mathbf{r}$, which in the sequel will denote the position vector in the physical space. We restrict ourselves to smooth phase space functions, which form a Lie algebra, given by the Poisson bracket $\{\cdot, \cdot\}$, defined as

$$\{a_1, a_2\} = \sum_{i=1}^{N} \left(\frac{\partial a_1}{\partial \mathbf{q}_i} \frac{\partial a_2}{\partial \mathbf{p}_i} - \frac{\partial a_1}{\partial \mathbf{p}_i} \frac{\partial a_2}{\partial \mathbf{q}_i} \right).$$

From Eq. (2.1), the time evolution equation for $a(\mathbf{Q}, \mathbf{P}, \mathbf{r}, t)$ can be written in the form

$$\partial_t a(\mathbf{Q}, \mathbf{P}, \mathbf{r}, t) = \{a, H\}, \tag{2.2}$$

where ∂_t denotes the partial derivative with respect to time. The values of $(\mathbf{Q}(t), \mathbf{P}(t))$ at time t, obtained by solving Eq. (2.1) for given initial values $(\mathbf{Q}, \mathbf{P})$, are expressed by the following canonical transformations [1]:

$$(\mathbf{Q}(t), \mathbf{P}(t)) = S^t (\mathbf{Q}, \mathbf{P}), \quad \text{with } S^t = e^{\mathscr{L}t}, \tag{2.3}$$

where $\mathscr{L}[\cdot] = \{\cdot, H\}$ is the *Liouville operator*. The time-evolved phase function $a(\mathbf{Q}, \mathbf{P}, \mathbf{r}, t)$ can be computed from Eq. (2.2) and also by exploiting the canonical transformations (2.3). It attains the form

$$a(\mathbf{Q}, \mathbf{P}, \mathbf{r}, t) = e^{\mathscr{L}t} a(\mathbf{Q}, \mathbf{P}, \mathbf{r}) = a(e^{\mathscr{L}t}(\mathbf{Q}, \mathbf{P}), \mathbf{r}), \tag{2.4}$$

which holds under the assumption of smoothness of the phase functions $a(\mathbf{Q}, \mathbf{P}, \mathbf{r}, t)$. In other words, under a canonical transformation, a (smooth enough) dynamical function $a(\mathbf{Q}, \mathbf{P}, \mathbf{r})$ goes over the same function of its time-evolved arguments. At the macroscopic level, the relevant quantities are represented by continuous functions in the physical space–time, which we denote by $A(\mathbf{r}, t)$. One of the basic tenets of statistical mechanics claims that for any observable $a(\mathbf{Q}, \mathbf{P}, \mathbf{r}, t)$, there exists a unique macroscopic function $A(\mathbf{r}, t)$, given by the *ensemble average* of $a(\mathbf{Q}, \mathbf{P}, \mathbf{r}, t)$, with respect to the density $F(\mathbf{Q}, \mathbf{P})$:

$$A(\mathbf{r}, t) = \langle a \rangle = \int a(\mathbf{Q}, \mathbf{P}, \mathbf{r}, t) F(\mathbf{Q}, \mathbf{P}) d\mathbf{Q} d\mathbf{P}. \tag{2.5}$$

It is worth remarking that the converse does not hold. There exist, in fact, certain thermal quantities, which typically play an important role in thermodynamics, that do not enjoy a mechanical definition and cannot be defined as shown in Eq. (2.5). For instance, one such quantity is the entropy, which is not a property of a single particle, but rather describes the overall "state of disorder" of the system. In this case, the corresponding dynamical function $a(\mathbf{Q}, \mathbf{P}, \mathbf{r}, t)$ is not a given, fixed microscopic dynamical function, but it depends on the probability density $F(\mathbf{Q}, \mathbf{P})$ [1]. Using Eq. (2.4), we can rewrite Eq. (2.5) as

$$A(\mathbf{r}, t) = \int \left[e^{\mathscr{L}t} a(\mathbf{Q}, \mathbf{P}, \mathbf{r}) \right] F(\mathbf{Q}, \mathbf{P}) d\mathbf{Q} d\mathbf{P}, \qquad (2.6)$$

which therefore expresses the law of motion of a macroscopic quantity as induced by the Hamiltonian dynamics of the corresponding observable in the phase space. Equation (2.6) suggests an analogy with the Heisenberg representation of quantum mechanics: the average is computed by keeping the state of the system fixed and by letting the dynamical phase space variables evolve in time. Conversely, in the Schrödinger-like picture, one may transfer the time dependence from the microscopic dynamical function to the probability density by employing the time-evolved density

$$F(\mathbf{Q}, \mathbf{P}, t) = e^{-\mathscr{L}t} F(\mathbf{Q}, \mathbf{P}).$$

This allows us to rewrite Eq. (2.6) in the form

$$A(\mathbf{r}, t) = \int a(\mathbf{Q}, \mathbf{P}, \mathbf{r}) F(\mathbf{Q}, \mathbf{P}, t) d\mathbf{Q} d\mathbf{P}, \qquad (2.7)$$

where the phase space density F obeys the celebrated Liouville equation

$$\partial_t F(\mathbf{Q}, \mathbf{P}, t) = -\mathscr{L} F(\mathbf{Q}, \mathbf{P}, t). \qquad (2.8)$$

The solutions of Eq. (2.8) are time-independent densities satisfying the relation

$$\{H, F_{eq}\} = 0. \qquad (2.9)$$

Every distribution of the form $F_{eq} = F_{eq}(H(\mathbf{Q}, \mathbf{P}))$ is readily seen to be a solution of Eq. (2.9). We mention in particular the density pertaining to the equilibrium canonical ensemble:

$$F_{eq}(\mathbf{Q}, \mathbf{P}) = \frac{1}{Z} e^{-\beta H(\mathbf{Q}, \mathbf{P})}, \qquad (2.10)$$

where $\beta = 1/(k_B T)$, k_B is Boltzmann's constant, T is the temperature of the particle system, kept fixed by an external heat bath, and Z denotes the canonical partition function, which is related to the thermodynamic Helmholtz free energy [2]. In spite of its conceptual relevance, the Liouville equation (2.8) is intractable from a practical point of view, due to the large number of particles. Thus, a more efficient reduced description can be obtained by projecting Eq. (2.8) onto a subspace of the whole phase space. To this end, we introduce the shorthand notation $\mathbf{z}_i = (\mathbf{q}_i, \mathbf{p}_i)$. We define the s-particle marginals $F_s(\mathbf{z}_1, \ldots, \mathbf{z}_s, t)$ as

$$F_s(\mathbf{z}_1, \ldots, \mathbf{z}_s, t) = \int F(\mathbf{z}_1, \ldots, \mathbf{z}_N, t) \prod_{j=s+1}^{N} d\mathbf{z}_j, \quad \text{with } j = 1, \ldots, N, \quad (2.11)$$

which denotes the probability density of the first s particles (or of any given group of s particles) at time t. To find an equation for the marginal F_s, we integrate Eq. (2.8) with respect to $d\mathbf{z}_{s+1}, \ldots, d\mathbf{z}_N$, thus obtaining the so-called BBGKY hierarchy [3], which reads

$$\partial_t F_s = \mathcal{L}_s F_s + \mathcal{C}_{s+1} F_{s+1}, \quad \text{with } s = 1, \ldots, N,$$

where

$$\mathcal{L}_s F_s = -\sum_{i=1}^{s} \mathbf{p}_i \cdot \frac{\partial F_s}{\partial \mathbf{q}_i} + \sum_{i=1}^{s} \sum_{j=1}^{s} \frac{\partial \Phi(|\mathbf{q}_i - \mathbf{q}_j|)}{\partial \mathbf{q}_i} \cdot \frac{\partial F_s}{\partial \mathbf{p}_i},$$

$$\mathcal{C}_{s+1} F_{s+1} = (N - s) \sum_{i=1}^{s} \int d\mathbf{z}_{s+1} \frac{\partial \Phi(|\mathbf{q}_i - \mathbf{q}_{s+1}|)}{\partial \mathbf{q}_i} \cdot \frac{\partial F_{s+1}}{\partial \mathbf{p}_i}. \quad (2.12)$$

The meaning of Eq. (2.12) is clear: the variation in time of F_s depends on the interaction of the s particles among themselves (which is the contribution provided by the operator $\mathcal{L}_s$) and on the interaction of the first s particles with the rest of the system (represented by the operator $\mathcal{C}_{s+1}$).

Let us also inspect the microscopic dynamical functions more closely. We introduce the one-point *empirical distribution* of the many-particle system, defined as

$$\sum_{i=1}^{N} \delta(\mathbf{r} - \mathbf{q}_i)\delta(\boldsymbol{\xi} - \mathbf{p}_i), \quad (2.13)$$

where $\mathbf{r}$ and $\boldsymbol{\xi}$ are parameters corresponding, respectively, to the position and momentum vectors in the single-particle space. In Sect. 2.2, we will clarify, in particular, in what sense the expression (2.13) is "well represented" by the distribution function entering the Boltzmann equation. Here it suffices to notice that the marginal F_1 is related to the phase space average (2.7) of the one-point empirical distribution (2.13), as follows:

$$\left\langle \sum_{i=1}^{N} \delta(\mathbf{r} - \mathbf{q}_i)\delta(\boldsymbol{\xi} - \mathbf{p}_i) \right\rangle = N F_1(\mathbf{r}, \boldsymbol{\xi}, t). \quad (2.14)$$

Furthermore, as we will see in Chap. 4, many phase functions of interest are obtained by multiplying the phase space distribution (2.13) by some function of $\boldsymbol{\xi}$ (see, e.g., Eq. (2.44)) and by integrating over the momentum space [4]. For instance, the local number density $n(\mathbf{Q}, \mathbf{r})$ reads

$$n(\mathbf{Q}, \mathbf{r}) = \sum_{i=1}^{N} \delta(\mathbf{r} - \mathbf{q}_i). \tag{2.15}$$

The macroscopic number density $n(\mathbf{r}, t)$, which, as discussed in Chap. 3, results from a proper integration of the density F_1 in the single-particle space, can also be derived by taking the ensemble average of the highly discontinuous phase function (2.15) with respect to the phase space density $F(\mathbf{Q}, \mathbf{P}, t)$. Analogously, we define the two-point empirical density as

$$\sum_{i=1}^{N} \sum_{\substack{j=1 \\ j \neq i}}^{N} \delta(\mathbf{r} - \mathbf{q}_i)\delta(\boldsymbol{\xi} - \mathbf{p}_i)\delta(\mathbf{r}_1 - \mathbf{q}_j)\delta(\boldsymbol{\xi}_1 - \mathbf{p}_j), \tag{2.16}$$

whose phase space average yields $N(N-1)F_2(\mathbf{r}, \boldsymbol{\xi}, \mathbf{r}_1, \boldsymbol{\xi}_1, t)$. An integration over the momenta leads to the two-point density function $n(\mathbf{r}, \mathbf{r}_1, t)$, which typically differs from the product of the one-point densities $n(\mathbf{r}, t)n(\mathbf{r}_1, t)$ as long as the distance $|\mathbf{r}-\mathbf{r}_1|$ is small. The deviation, which stems from the presence of particle interactions, corresponds to the *two-point correlation* [1]. In particular, if we denote by V the volume of the system, then the two-point function $n(\mathbf{r}, \mathbf{r}_1)$ for a homogeneous system can be written as

$$n(\mathbf{r}, \mathbf{r}_1) = \left(\frac{N}{V}\right)^2 g^{(2)}(\mathbf{r}, \mathbf{r}_1), \tag{2.17}$$

where the correlation function $g^{(2)}(\mathbf{r}, \mathbf{r}_1)$ measures the extent to which the structure of a fluid deviates from complete randomness. If the system is also isotropic, then $g^{(2)}$ depends only on the magnitude of the separation $s = |\mathbf{r} - \mathbf{r}_1|$: in this case, the pair correlation function is denoted by $g(s)$ and is referred to as the *radial distribution function* [5]. By taking the ensemble average of Eq. (2.16), in which the phase space coordinates $(\mathbf{q}_i, \mathbf{p}_i)$ and $(\mathbf{q}_j, \mathbf{p}_j)$ are evaluated at two distinct times, say t and t_1,

$$\sum_{i=1}^{N} \sum_{j=1}^{N} \delta(\mathbf{r} - \mathbf{q}_i(t))\delta(\boldsymbol{\xi} - \mathbf{p}_i(t))\delta(\mathbf{r}_1 - \mathbf{q}_j(t_1))\delta(\boldsymbol{\xi}_1 - \mathbf{p}_j(t_1)), \tag{2.18}$$

one obtains the two-time joint probability distribution [1]

$$f_2(\mathbf{r}, \boldsymbol{\xi}, t; \mathbf{r}_1, \boldsymbol{\xi}_1, t_1) = N^2 F_2(\mathbf{r}, \boldsymbol{\xi}, t; \mathbf{r}_1, \boldsymbol{\xi}_1, t_1), \tag{2.19}$$

which allows the computation of time correlation functions, whose properties will be explored in Chap. 4.

Before concluding this section, it is instructive to outline some aspects concerning the derivation of the Boltzmann equation for a hard-sphere gas from the hierarchy (2.12). This can be done by taking a proper limit of the time evolution equation for the marginal F_1. To this end, it is common practice to replace the momentum $\boldsymbol{\xi}$ with the velocity $\mathbf{v}$. We denote, then, by $\mathbf{v}, \mathbf{v}_1, \mathbf{v}'$, and $\mathbf{v}'_1$ the precollisional velocities of the two particles in respectively the *direct* and *inverse* encounters (the explicit relations between postcollisional and precollisional velocities will be made clear in Sect. 2.2). Let $\mathbf{r}$ be the position of the center of the target particle equipped with velocity $\mathbf{v}$. In a reference frame centered on the target particle, the latter is at rest and is endowed with twice the actual diameter δ. The colliding particle, which is regarded as a point mass equipped with an incoming velocity $\mathbf{v}_1 - \mathbf{v}$, hits the so-called *protection* sphere [6] at the point $\mathbf{r}_1 = \delta\mathbf{n}$, where $\mathbf{n}$ is a unit vector, and is contained within a volume $\delta^2 d\mathbf{n}|(\mathbf{v}_1 - \mathbf{v}) \cdot \mathbf{n}|dt$. The time evolution equation for F_1 attains the form [3, 6]

$$\frac{\partial F_1}{\partial t} + \mathbf{v} \cdot \frac{\partial F_1}{\partial \mathbf{r}} = (N-1)\delta^2 \int\limits_{\mathbb{R}^3} \int\limits_{S_-} \left[F_2(\mathbf{r}, \mathbf{v}', \mathbf{r} - \delta\mathbf{n}, \mathbf{v}'_1, t) - F_2(\mathbf{r}, \mathbf{v}, \mathbf{r} + \delta\mathbf{n}, \mathbf{v}_1, t) \right]$$

$$\times |(\mathbf{v}_1 - \mathbf{v}) \cdot \mathbf{n}| d\mathbf{n} d\mathbf{v}_1, \tag{2.20}$$

where the integration is over the hemisphere $S_- = \{\mathbf{n} \in S^2 | (\mathbf{v}_1 - \mathbf{v}) \cdot \mathbf{n} < 0\}$. It is worth noticing that the probability densities F_2 in (2.20) are both expressed in terms of the precollisional velocities pertaining respectively to the inverse and the direct encounters. Here, then, is the Boltzmann argument. For a rarefied gas contained in a box with volume $1\,\mathrm{cm}^3$ at room temperature and atmospheric pressure, we have $N \simeq 10^{20}$ and $\delta = 10^{-8}\,\mathrm{cm}$, so that $(N-1)\delta^2 \simeq N\delta^2 = 1\,\mathrm{m}^2$; cf. [3]. The limit $N \to \infty$, $\delta \to 0$, with $N\delta^2$ finite, is the so-called Boltzmann–Grad limit; see also Sect. 2.3. In this limit, if each F_s tends to a limit and this limit is sufficiently smooth, the BBGKY hierarchy (2.12) transforms into the so-called Boltzmann hierarchy [3]. Furthermore, since the volume occupied by the particles corresponds to about $N\delta^3 \simeq 10^{-4}\,\mathrm{cm}^3$, the collision between two particles is a rather rare event. Thus, the two colliding particles may be thought of as completely uncorrelated *before* the collision, and their joint probability density may be factorized (assumption of *molecular chaos*, or *Stosszahlansatz*) as

$$F_2(\mathbf{r}, \mathbf{v}, \mathbf{r}_1, \mathbf{v}_1, t) = F_1(\mathbf{r}, \mathbf{v}, t) F_1(\mathbf{r}_1, \mathbf{v}_1, t). \tag{2.21}$$

Taking into account the remarks above, one obtains the Boltzmann equation for hard spheres

$$\frac{\partial F_1}{\partial t} + \mathbf{v} \cdot \frac{\partial F_1}{\partial \mathbf{r}} = N\delta^2 \int\limits_{\mathbb{R}^3} \int\limits_{S_-} \left[F_1(\mathbf{r}, \mathbf{v}', t) F_1(\mathbf{r}, \mathbf{v}'_1, t) - F_1(\mathbf{r}, \mathbf{v}, t) F_1(\mathbf{r}, \mathbf{v}_1, t) \right] |(\mathbf{v}_1 - \mathbf{v}) \cdot \mathbf{n}| d\mathbf{n} d\mathbf{v}_1.$$

$$\tag{2.22}$$

We are not going to dwell further here on the rigorous derivation of the Boltzmann equation, which is beyond the scope of this work. Rather, we wish to outline the heuristic argument that guided Ludwig Boltzmann in the derivation of the celebrated kinetic equation bearing his name.

2.2 The Boltzmann Equation

The Boltzmann equation is an evolution equation for the probability density of a given particle, which we denote by $f(\mathbf{r}, \mathbf{v}, t)$. This quantity should not be confused in principle with the fraction of molecules located in the cell of size $d\mathbf{r} \times d\mathbf{v}$ around the point $\mathbf{r}, \mathbf{v}$. Although there is no a priori relationship between the two concepts, it can be shown that they are closely related. Let us reformulate this problem in the following manner. As in Sect. 2.1, let us denote by $\mathbf{z}$ the pair $(\mathbf{r}, \mathbf{v})$. Then, given a system of N identical particles whose physical state is given by the sequence $\mathbf{z}_1 \ldots \mathbf{z}_N$, we denote by $F_\Delta(\mathbf{z}_1, \ldots, \mathbf{z}_N)$ the fraction of particles localized in a volume Δ of the single-particle space, i.e.,

$$F_\Delta(\mathbf{z}_1, \ldots, \mathbf{z}_N) = \frac{1}{N} \sum_{i=1}^{N} \chi_\Delta(\mathbf{z}_i) = \int_\Delta \mu(d\mathbf{z}), \qquad (2.23)$$

where χ_Δ is the indicator function and

$$\mu(d\mathbf{z}) = \frac{1}{N} \sum_{i=1}^{N} \delta(\mathbf{z} - \mathbf{z}_i) d\mathbf{z}$$

is a measure in the single-particle space. In the definition of $\mu(d\mathbf{z})$ in Eq. (2.2), the reader will also recognize the structure of the empirical distribution, defined in Eq. (2.13). Thus, one may wonder in which sense the random variable $F_\Delta(\mathbf{z}_1, \ldots, \mathbf{z}_N)$ in Eq. (2.23) can be reasonably approximated by

$$\int_\Delta f(\mathbf{z}, t) d\mathbf{z}.$$

It is possible to prove [3] that if $\mathbf{z}_1, \ldots, \mathbf{z}_N$ are independent identically distributed random variables with law $f(\mathbf{z}, t)$, then $\mu(d\mathbf{z})$ is close to $f(\mathbf{z}, t) d\mathbf{z}$, in a sense specified by the law of large numbers, when N diverges. That is, the Boltzmann equation refers to the limiting situation such that the many-particle system specified by the sequence $\mathbf{z}_1, \ldots, \mathbf{z}_N$ can be actually considered a large number of copies of independent single-particle systems for which $\mu(d\mathbf{z})$ and $f(\mathbf{z})d\mathbf{z}$ can be regarded as the same object. Moreover, it is common to rescale $f(\mathbf{z}, t)$ by setting

$$f(\mathbf{r}, \mathbf{v}, t) = N F_1(\mathbf{r}, \mathbf{v}, t), \qquad (2.24)$$

cf. Eq. (2.14). The quantity $f(\mathbf{r}, \mathbf{v}, t) d\mathbf{r} d\mathbf{v}$ is to be regarded as the *average* number of particles contained in the element $d\mathbf{r} d\mathbf{v}$ at a given time t, when the fluctuations which occur in a short time interval dt are neglected [7]. The definition of the function f relies, hence, on probability concepts, and hereinafter, we will tacitly assume that the macroscopic variables computed from the knowledge of the distribution function f at a certain time t result from an average over the time interval dt around t. We will return in more detail to this issue in Sect. 3.1. It is worth mentioning that the role of fluctuations in the single-particle space was the subject of a research investigation [8–10] that led to the construction of a fluctuating Boltzmann equation that recovers the standard Boltzmann equation on average and from which the macroscopic equations of fluctuating hydrodynamics [11] can be derived.

Let us now outline some basic aspects concerning the geometry of an elastic collision between two particles endowed with equal mass m.

We consider the scattering of a particle, labeled 2, equipped with velocity $\mathbf{v}_1$, induced by the *target* particle labeled 1, with velocity $\mathbf{v}$. The pre- and postcollisional velocities are related by momentum and kinetic energy conservation:

$$\mathbf{v} + \mathbf{v}_1 = \mathbf{v}' + \mathbf{v}_1',$$
$$\mathbf{v}^2 + \mathbf{v}_1^2 = \mathbf{v}'^2 + \mathbf{v}_1'^2. \tag{2.25}$$

Introducing the center-of-mass variables

$$\mathbf{G} = \frac{1}{2}(\mathbf{v} + \mathbf{v}_1),$$
$$\mathbf{g} = \mathbf{v}_1 - \mathbf{v},$$

and the corresponding variables $\mathbf{G}'$ and $\mathbf{g}'$ for the postcollisional velocities (in the direct encounter), the conservation equations (2.25) yield $g = g'$, where $g \equiv |\mathbf{g}|$. Hence, the relative velocity is changed by the collision only in direction and not in magnitude.

We consider, then, a reference frame in which the target particle is at rest. In this frame, the geometry of the scattering is determined by the collision (unit) vector $\mathbf{n}$, which forms an angle θ, called the collision angle, with the direction of $\mathbf{g}$ and halves the angle between $\mathbf{g}$ and $\mathbf{g}'$; cf. Fig. 2.1. Therefore, in the direct encounter, the precollisional and postcollisional velocities, respectively $\mathbf{g}$ and $\mathbf{g}'$, are related by

$$\mathbf{g} - \mathbf{g}' = 2(\mathbf{g} \cdot \mathbf{n})\mathbf{n}. \tag{2.26}$$

Using (2.26), it is possible to derive the following relations between pre- and post-collisional velocities:

$$\mathbf{v}' - \mathbf{v} = (\mathbf{g} \cdot \mathbf{n})\mathbf{n},$$
$$\mathbf{v}_1' - \mathbf{v}_1 = -(\mathbf{g} \cdot \mathbf{n})\mathbf{n}. \tag{2.27}$$

the relations (2.27) show that the dynamical effect of the encounter is known when the direction of the unit vector $\mathbf{n}$ is also determined. Yet the latter vector cannot be computed by relying only on the balance equation (2.25), because there are two

Fig. 2.1 Scattering kinematics for hard *spheres*, with precollisional relative velocity **g**, postcollisional velocity **g**$'$, impact parameter b, collision angle θ, and collision vector **n**

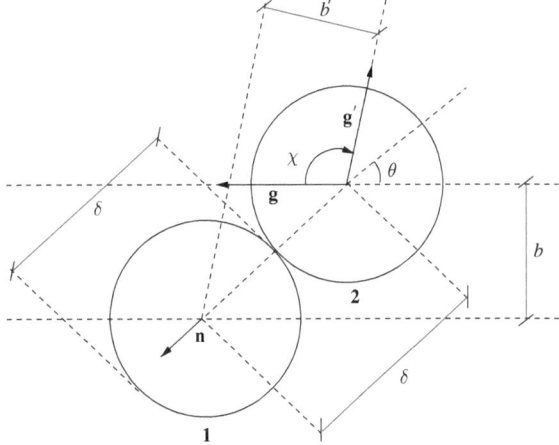

Fig. 2.2 Collision *cylinder* for particles with relative velocity **g** that collide with the target particle during a time interval dt .

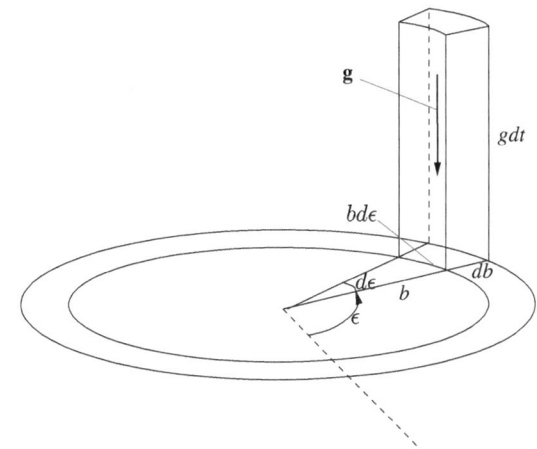

geometric variables to be specified. The first is the azimuthal angle ε, describing the orientation of the plane containing **g** and **g**$'$ (which lie in the same plane, as specified below) with a reference plane; cf. Fig. 2.2.

The second geometric variable of the encounter is the impact parameter b, which corresponds to the perpendicular distance between the center of the target particle 1 and the original line of motion of the scattered particle 2, identified by the relative velocity **g**, as shown in Fig. 2.1. Moreover, the angle of deflection $\chi = \pi - 2\theta$ is, in general, a function of g and of b, with a specific functional relation $\chi(g, b)$ that depends on the particle interaction potential (see below). Analogously, one may also introduce the postcollisional impact parameter b', defined with respect to the velocity **g**$'$, as illustrated in Fig. 2.1. The conservation of the angular momentum entails that the collision takes place in a plane, called the *collision plane*, and yields $b = b'$.

Let us now focus on the statistics of the encounters. To this end, we use, for simplicity, the abbreviations $f \equiv f(\mathbf{r}, \mathbf{v}, t)$, $f_1 \equiv f(\mathbf{r}, \mathbf{v}_1, t)$, $f' \equiv f(\mathbf{r}, \mathbf{v}', t)$, and $f_1' \equiv f(\mathbf{r}, \mathbf{v}_1', t)$. The collision between the target particle 1 and an arbitrary particle 2 occurs if the latter is located within the cylinder with volume $dV = (g dt)(b db d\varepsilon)$ shown in Fig. 2.2. The (average) number of the particles 2 equipped with a velocity in the range $[\mathbf{v}_1, \mathbf{v}_1 + d\mathbf{v}_1]$ and located in any such cylinders is given by $f_1 d\mathbf{v}_1 dV$. We can imagine such a cylinder to be associated with any of the particles 1, so that the total number of encounters, under the assumption of molecular chaos considered by Boltzmann, is given by

$$f \, d\mathbf{v} d\mathbf{r} \, f_1 d\mathbf{v}_1 (g dt)(b db d\varepsilon). \tag{2.28}$$

The surface element $b db d\varepsilon$ in Eq. (2.28) can be written in terms of the element $d\mathbf{n} = \sin \theta d\theta d\varepsilon$, using the relation

$$b db d\varepsilon = \sigma(g, b) \, d\mathbf{n}. \tag{2.29}$$

The quantity $\sigma(g, b)$ in Eq. (2.29) is called the differential scattering cross section and is defined by the relation

$$\sigma(g, b) = \frac{b}{\sin \theta} \left| \frac{db}{d\theta} \right|. \tag{2.30}$$

It contains all the detailed information on the particle interactions [3, 6, 7]. In particular, by assuming, as above, that the interaction potential $\Phi(s)$ depends only on the magnitude of the relative distance $s = |\mathbf{r} - \mathbf{r}_1|$ between the particles, the specific functional relation among θ, g, and b can be obtained from the general relation

$$\theta(g, b) = \int_0^{z^*} \frac{dz}{\sqrt{1 - z^2 - \frac{4\Phi(b/z)}{mg^2}}}, \quad \text{with } 1 - z^{*2} - \frac{4\Phi(b/z^*)}{mg^2} = 0. \tag{2.31}$$

Here $z = b/s$ is a dimensionless inverse distance, and z^* denotes its value at the minimum particle distance s_{min}. Equation (2.31) provides the fundamental relationship $\theta(b, g)$. Inverting this relation, one obtains $b(\theta, g)$, which thus yields $\sigma(g, \theta)$ as a function of g and θ. In particular, for hard spheres of diameter δ, as shown in Fig. 2.1, the relation $b = \delta \sin \theta$ holds. This yields

$$\sigma_{HS}(g, \theta) = \delta^2 \cos \theta, \tag{2.32}$$

which hence shows that for a hard-sphere gas, the scattering cross section depends only on θ. On the other hand, for power law potentials of the form $\Phi(s) = \Phi_0 s^{(1-n)}$, with Φ_0 and n constants, we have

$$\sigma_n(g, \theta) = \beta(\theta) g^{-\frac{4}{n-1}}, \tag{2.33}$$

where $\beta(\theta)$ is a function of θ alone [6]. The case $n = 5$ occupies a distinct place in the Boltzmann equation theory, since it allows a considerable simplification: in this case, the product $g\sigma(g, \theta)$, which enters the expression of the collision operator (to be made explicit below, cf. Eq. (2.41)) depends only on θ. This result was discovered by Maxwell, whence the fictitious molecules interacting via such a purely repulsive potential are called Maxwell molecules [6]. We will return to Maxwell molecules in Chap. 5, to discuss the derivation of hydrodynamics from kinetic models of the Boltzmann equation.

In the inverse encounters, characterized by pre- and postcollisional velocities given, respectively, by $\mathbf{v}'$, $\mathbf{v}'_1$ and $\mathbf{v}$, $\mathbf{v}_1$, the differential scattering cross section is a function of g' and b'. Using the relations $g = g'$ and $b = b'$, one immediately obtains

$$\sigma\left(g', b'\right) = \sigma\left(g, b\right), \tag{2.34}$$

which reflects the time reversibility of the collisional event. In particular, we recall that the scattering cross section σ is related to the conditional probability density $W\left(\mathbf{v}', \mathbf{v}'_1 | \mathbf{v}, \mathbf{v}_1\right)$ of a collision that, in the direct encounter, carries the two particles from the velocities $\mathbf{v}$ and $\mathbf{v}_1$ into the velocities $\mathbf{v}'$ and $\mathbf{v}'_1$ [6, 12]. The explicit relation between W and σ reads

$$W\left(\mathbf{v}', \mathbf{v}'_1 | \mathbf{v}, \mathbf{v}_1\right) = 2m^4 \frac{\sigma(g, \theta)}{\cos \theta} \delta\left(\mathbf{v} + \mathbf{v}_1 - \mathbf{v}' - \mathbf{v}'_1\right) \delta\left(\mathbf{v}^2 + \mathbf{v}_1^2 - \mathbf{v}'^2 - \mathbf{v}'^2_1\right), \tag{2.35}$$

where the two delta functions ensure the conservation of the momentum and of the kinetic energy. Moreover, from (2.34) and (2.35), it follows that

$$W\left(\mathbf{v}', \mathbf{v}'_1 | \mathbf{v}, \mathbf{v}_1\right) = W\left(\mathbf{v}, \mathbf{v}_1 | \mathbf{v}', \mathbf{v}'_1\right), \tag{2.36}$$

which is another expression of the time reversibility of the collisional event. The conditional probability density is also symmetric under the exchange of particles

$$W\left(\mathbf{v}, \mathbf{v}_1 | \mathbf{v}', \mathbf{v}'_1\right) = W\left(\mathbf{v}_1, \mathbf{v} | \mathbf{v}'_1, \mathbf{v}'\right). \tag{2.37}$$

Let us now consider the set of particles located within $d\mathbf{r}$ and with velocities in the range $[\mathbf{v}, \mathbf{v} + d\mathbf{v}]$, and let us also denote by L and G the negative and positive contributions to the variation of f due to the collisions. The loss term L can be computed by considering the direct encounter in which the precollisional velocities $\mathbf{v}$, $\mathbf{v}_1$ transform, according to Eq. (2.27), into the postcollisional ones. Therefore, using (2.28), one can write

$$L = f f_1 g \sigma(g, b) \mathbf{dn} \mathbf{dv} \mathbf{dv}_1 \mathbf{dr} dt. \tag{2.38}$$

The gain term G can be found in a similar manner by considering inverse encounters. The relation between pre- and postcollisional velocities in the inverse encounter follows directly from Eq. (2.27) on interchanging $\mathbf{v}$, $\mathbf{v}_1$ and $\mathbf{v}'$, $\mathbf{v}'_1$, and by replacing $\mathbf{n}$ by $(-\mathbf{n})$ and $\mathbf{g}$ by $\mathbf{g}'$ [7]. Thus, one obtains

$$G = f' f'_1 g' \sigma(g', b') \mathbf{dn} \mathbf{dv}' \mathbf{dv}'_1 \mathbf{dr} dt. \tag{2.39}$$

Since the Jacobian of the transformation from precollisional to postcollisional velocities is unitary, i.e., $\mathbf{dv} \mathbf{dv}_1 = \mathbf{dv}' \mathbf{dv}'_1$, and because of Eq. (2.34), we can use $g \sigma(g, b) \mathbf{dv} \mathbf{dv}_1$ also in (2.39). Dividing both terms L and G by $\mathbf{dv} \mathbf{dr} dt$ and performing the integration over the variables $\mathbf{v}_1$ and $\mathbf{n}$, we obtain the Boltzmann equation in absence of external forces:

$$\partial_t f(\mathbf{r}, \mathbf{v}, t) + \mathbf{v} \cdot \frac{\partial f}{\partial \mathbf{r}} f = Q(f, f), \tag{2.40}$$

where

$$Q(f, f) = \int \int \left(f' f'_1 - f f_1 \right) g \sigma(g, \theta) \mathbf{dn} \mathbf{dv}_1 \tag{2.41}$$

denotes a nonlinear integral collision operator. In particular, using the expression (2.32) of the scattering cross section for hard spheres, one readily recovers Eq. (2.22). We recall that the integration over $\mathbf{n}$ corresponds to a double integration over $\varepsilon = [0, 2\pi]$ and $\theta = [0, \pi/2]$; cf. Fig. 2.1. Alternatively, the collision operator can be expressed in terms of the conditional probability densities introduced in Eq. (2.35), and using (2.36), it attains the form

$$Q(f, f) = \int \int \int \left(f' f'_1 - f f_1 \right) W \left(\mathbf{v}, \mathbf{v}_1 | \mathbf{v}', \mathbf{v}'_1 \right) \mathbf{dv}_1 \mathbf{dv}' \mathbf{dv}'_1, \tag{2.42}$$

which is sometimes referred to as the *quasichemical* representation.
A basic property of the operator (2.41), which will be employed in the sequel of this work, concerns the conservation of mass, momentum, and kinetic energy of the particles during the collisions. Defining the vector of the so-called *elementary collision invariants* as

$$\psi(\mathbf{v}) = \left[\Psi_1, \psi_2, \Psi_3 \right], \tag{2.43}$$

with

$$\Psi_1 = m, \quad \psi_2 = m\mathbf{v}, \quad \Psi_3 = \frac{m}{2} v^2, \tag{2.44}$$

and $v \equiv |\mathbf{v}|$, one obtains the important relation

$$\int Q(f, f) \psi(\mathbf{v}) \mathbf{dv} = \mathbf{0}, \tag{2.45}$$

which expresses the conservation of the mean values of mass, momentum, and kinetic energy of the particles during the collisions; cf. Sect. 3.1. In general, any function $\Phi(\mathbf{r}, \mathbf{v}, t)$ satisfying the relation

$$\phi_1' + \phi' = \phi_1 + \phi \tag{2.46}$$

is a collision invariant and can be written as a linear combination of the elements of $\psi(\mathbf{v})$.

2.2.1 H-Theorem and the Maxwellian Distribution

Let us consider a uniform gas in the absence of external forces. In this case, the Boltzmann equation (2.40) reads

$$\partial_t f(\mathbf{r}, \mathbf{v}, t) = \int\int (f' f_1' - f f_1) g \sigma(g, \theta) \mathrm{d}\mathbf{n} \mathrm{d}\mathbf{v}_1. \tag{2.47}$$

Furthermore, let H be the complete integral (i.e., the integral over all values of velocities) defined by the equation

$$H(t) = \int f \log f \mathrm{d}\mathbf{v}. \tag{2.48}$$

By computing the time derivative of the H-function defined in Eq. (2.48) through Eq. (2.47), one obtains

$$\partial_t H(t) = \frac{1}{4} \int\int\int g\sigma(f' f_1' - f f_1) \ln\left(\frac{f f_1}{f' f_1'}\right) \mathrm{d}\mathbf{v} \mathrm{d}\mathbf{v}_1 \mathrm{d}\mathbf{n}. \tag{2.49}$$

The structure of Eq. (2.49) leads to the following inequality, called the *H-theorem*:

$$\partial_t H(t) \leq 0. \tag{2.50}$$

Equation (2.50) shows that the function $H(t)$ decreases monotonically over time. Consequently, the particle system keeps no track of its initial distribution, and is driven toward a final state described by the distribution f^{GM}. The latter satisfies the relation

$$f^{\mathrm{GM}}(\mathbf{v}) f^{\mathrm{GM}}(\mathbf{v}_1) = f^{\mathrm{GM}}(\mathbf{v}') f^{\mathrm{GM}}(\mathbf{v}_1'). \tag{2.51}$$

Taking the logarithm of both sides of (2.51), we obtain a relation of the form (2.46): $\ln f^{GM}$ is a collision invariant and hence can be written as a sum of the elementary collision invariants (2.44):

$$\ln f^{GM}(\mathbf{v}) = \alpha_1 + \alpha_2 \cdot m\mathbf{v} + \alpha_3 \frac{m}{2} v^2.$$

The coefficients $\alpha_1, \alpha_2, \alpha_3$ are determined by ensuring that the distribution f^{GM} yields the proper values for the mass density, momentum density, and internal energy density, denoted respectively by $\rho_0, \mathbf{u}_0, T_0$, with $\rho_0 = mn_0$:

$$\int f^{GM} d\mathbf{v} = \rho_0,$$

$$\int f^{GM} \mathbf{v} d\mathbf{v} = \rho_0 \mathbf{u}_0,$$

$$\frac{m}{2} \int f^{GM} (\mathbf{v} - \mathbf{u})^2 d\mathbf{v} = \frac{3}{2} n_0 k_B T_0. \tag{2.52}$$

Therefore, the general equilibrium distribution attains the form

$$f^{GM}(\mathbf{v}) = n_0 \left(\frac{m}{2\pi k_B T_0} \right)^{\frac{3}{2}} e^{-\frac{m(\mathbf{v} - \mathbf{u}_0)^2}{2k_B T_0}}. \tag{2.53}$$

If the macroscopic fields n, $\mathbf{u}$, T depend on $\mathbf{r}$ and t, the corresponding equilibrium distribution is then referred to as the *local Maxwellian*, and it reads

$$f^{LM}(\mathbf{r}, \mathbf{v}, t) = n(\mathbf{r}, t) \left(\frac{m}{2\pi k_B T} \right)^{\frac{3}{2}} e^{-\frac{m(\mathbf{v} - \mathbf{u}(\mathbf{r},t))^2}{2k_B T(\mathbf{r},t)}}. \tag{2.54}$$

The expression (2.54) represents an important reference distribution for describing a gas's behavior in close-to-equilibrium regimes.

2.3 Hydrodynamic Limit and Other Scalings

A point of remarkable interest in kinetic theory concerns the study of the scaling properties of the Boltzmann equation. In particular, one may investigate the structure of the solutions of the Boltzmann equation in the so-called *hydrodynamic limit*, which is obtained by taking the limits $N \to \infty$, $V \to \infty$, with N/V finite, where N is the total number of particles in a box Λ_ε with volume V. Following the approach given in [3], let us introduce a small parameter ε, and let the side of the box Λ_ε be proportional to ε^{-1}. Let $f^\varepsilon(\mathbf{r}, \mathbf{v}, t)$, with $\mathbf{r} \in \Lambda_\varepsilon$, be the number density of the particles in the box. Then we assume that the total number of particles is proportional to the volume of the box

$$\int_{\Lambda_\varepsilon \times \mathbb{R}^3} f^\varepsilon(\mathbf{r}, \mathbf{v}) d\mathbf{r} d\mathbf{v} = \varepsilon^{-3}, \tag{2.55}$$

and introduce the following scaling:

$$\hat{\mathbf{r}} = \varepsilon \mathbf{r}, \quad \tau = \varepsilon t, \tag{2.56}$$

with $\mathbf{r} \in \Lambda = [0, 1]$, and

$$\hat{f}(\hat{\mathbf{r}}, \mathbf{v}, \tau) = f^{\varepsilon}(\mathbf{r}, \mathbf{v}, t). \tag{2.57}$$

It is readily seen that the rescaled distribution $\hat{f}(\hat{\mathbf{r}}, \mathbf{v}, \tau)$ can suitably describe the particle system on the scale of the box and is normalized to unity:

$$\int_{\Lambda \times \mathbb{R}^3} \hat{f}(\hat{\mathbf{r}}, \mathbf{v}) d\hat{\mathbf{r}} d\mathbf{v} = 1. \tag{2.58}$$

As discussed in [3], while the description in terms of the distribution f^{ε} is called *microscopic*, the description given by the rescaled distribution $\hat{f}$ can be defined as *macroscopic*, since it provides a statistical description of the particle system on large length and time scales. From (2.56) and (2.57), the Boltzmann equation, in the absence of external forces, attains the structure

$$\partial_{\tau} \hat{f}(\hat{\mathbf{r}}, \mathbf{v}, \tau) + \mathbf{v} \cdot \nabla_{\hat{\mathbf{r}}} \hat{f} = \frac{1}{\varepsilon} Q(\hat{f}, \hat{f}), \tag{2.59}$$

which will be the starting point of the methods of reduced description treated in Chap. 3. We expend a few words now on another scaling, in order to clarify the nature of the Boltzmann–Grad limit, introduced earlier. As mentioned in Sect. 2.1, the Boltzmann–Grad limit corresponds to taking the limits $N \to \infty$, $\delta \to 0$, with $N\delta^2$ finite. If we now take $\varepsilon = \delta$, we require the particle number to be of order ε^{-2}, i.e., it goes with the power 2/3 of the volume:

$$\int_{\Lambda_{\varepsilon} \times \mathbb{R}^3} f^{\varepsilon}(\mathbf{r}, \mathbf{v}) d\mathbf{r} d\mathbf{v} = \varepsilon^{-2}. \tag{2.60}$$

Next, if we employ the scaling (2.56) and wish to keep the normalization of $\hat{f}(\hat{\mathbf{r}}, \mathbf{v}, \tau)$ to unity as in Eq. (2.58), we resort to a different scaling, given by

$$\hat{f}(\hat{\mathbf{r}}, \mathbf{v}, \tau) = \varepsilon^{-1} f^{\varepsilon}(\mathbf{r}, \mathbf{v}, t). \tag{2.61}$$

Hence, the Boltzmann equation for the distribution $\hat{f}$ reads

$$\partial_{\tau} \hat{f}(\hat{\mathbf{r}}, \mathbf{v}, \tau) + \mathbf{v} \cdot \nabla_{\hat{\mathbf{r}}} \hat{f} = Q(\hat{f}, \hat{f}). \tag{2.62}$$

Equation (2.62) is invariant under the scalings (2.56) and (2.61). It is also clear why the Boltzmann–Grad limit corresponds to a *low-density* limit: in this limit, in fact, the ratio of the particle number to the volume tends to zero [3]. The introduction of the parameter ε in the scaling (2.56) reflects the existence of different length and time scales in a gas. In particular, the structure of the hydrodynamic equations derived from the Boltzmann equation strongly depends on the magnitude of ε, which determines the appropriate gas dynamic regime.

As will also be discussed in Chap. 3, the presence of a definite time scale separation in an inhomogeneous gas allows us to construct the following representation of the relaxation to equilibrium: starting from an arbitrary nonequilibrium initial state, the sequence of collisions characterized by the time scale τ_{mf} (the mean time between collisions) triggers the evolution of the system toward the local equilibrium regime, which is reached in a "mesoscopic" time interval Δt. From then onward, one may assign locally a value to the macroscopic fields $\rho, \mathbf{u}, T$, which also then evolve according to a characteristic time scale τ_{macro}. Then in the final stage of the relaxation process, the slow dynamics of the macroscopic variables drives the particle system toward the final, homogeneous, equilibrium state (2.53).

2.4 Linearized Collision Integrals and Kinetic Models

One of the major shortcomings of the Boltzmann equation (2.40) concerns the nonlinear nature of the integral collision operator, which is traced back to the assumption of molecular chaos. The Maxwellian distributions introduced in Sect. 2.2.1 represent a first step in the description of a particle system: they describe equilibrium states characterized by the absence of dissipation. In going beyond equilibrium, one has to rely on approximation methods such as perturbation techniques: one typically expands the distribution function f in powers of the aforementioned parameter ε, which in certain favorable situations can be considered to be small. We focus here on the properties of the collision operator $Q(f, f)$ under such perturbation theory. Near global equilibrium, one can split the distribution function into the sum of two contributions:

$$f(\mathbf{r}, \mathbf{v}, t) = f^{\text{GM}}(\mathbf{v}) + \Delta f(\mathbf{r}, \mathbf{v}, t), \tag{2.63}$$

where $\Delta f(\mathbf{r}, \mathbf{v}, t) = f^{\text{GM}}(\mathbf{v}) h(\mathbf{r}, \mathbf{v}, t)$ represents a "small" deviation from the equilibrium distribution. By linearizing the Boltzmann collision operator around equilibrium, one obtains the *linearized* collision operator L,

$$Lh = 2\left(f^{\text{GM}}\right)^{-1} Q(f^{\text{GM}}, \Delta f) = \int f^{\text{GM}}(\mathbf{v}_1) \left(h'_1 + h' - h_1 - h\right) g\sigma(g, \theta) \mathrm{dn} \mathrm{d}\mathbf{v}_1, \tag{2.64}$$

where $\sigma(g, \theta)$ contains, as seen above, the details of the particle interactions. Then, by introducing the Hilbert space $\mathscr{H}$ endowed with scalar product $\langle g|h \rangle$ and norm $\|h\|$ defined by

$$\langle g|h\rangle = \frac{1}{n_0} \int f^{GM}(\mathbf{v})g(\mathbf{v})h(\mathbf{v})d\mathbf{v}, \quad \|h\|^2 = \langle h|h\rangle, \tag{2.65}$$

one finds that the operator L in Eq. (2.64) is self-adjoint, i.e.,

$$\langle g|Lh\rangle = \langle Lg|h\rangle.$$

Moreover, by setting $g = h$ in (2.64), one also obtains

$$\langle h|Lh\rangle = -\frac{1}{4}\int |h_1' + h' - h_1 - h)|^2 g\sigma(g,\theta)d\mathbf{n}d\mathbf{v}_1 d\mathbf{v} \leq 0, \tag{2.66}$$

where equality holds if and only if h is a collision invariant. A simple problem for the linearized operator L concerns the asymptotic behavior of a homogeneous gas, prepared in an initial state slightly deviating from equilibrium and described by a distribution of the form (2.63). In this case, one studies the linearized Boltzmann equation

$$\partial_t h = Lh. \tag{2.67}$$

If h is such that $\|h\|$ and $\langle h|Lh\rangle$ exist, then from (2.66), one obtains

$$\partial_t\left[\frac{1}{2}\|h\|^2\right] = \langle h|Lh\rangle \leq 0,$$

which entails that $\|h(t_2)\| \leq \|h(t_1)\|$ for $t_2 > t_1$, where equality holds if h is a collision invariant. Moreover, if f is chosen to have the same density, bulk velocity, and temperature of f^{GM} as in Eqs. (2.52), then

$$\langle \psi|h\rangle = 0, \tag{2.68}$$

and the only collision invariant that satisfies these relations is $h = 0$. Therefore, $\|h(t)\|$ decreases until it vanishes for $t \to \infty$. We also mention here that another remarkable problem concerns the study of the spectrum of L [6], i.e., the set of eigenvalues λ of L for which $(L - \lambda I)^{-1}$ is not a bounded operator in $\mathscr{H}$ or is not uniquely determined. This aspect will be elaborated in Chap. 5.

A complementary strategy to simplify the complicated structure of the Boltzmann equation consists in providing simpler expressions for the collision operator, which, in spite of neglecting many details about the particle interactions, allow one to retain the basic properties of the original collision operator. This is the philosophy behind the wealth of kinetic models proposed in the literature. The simplest example is represented by the BGK model [13–15], which reads

$$\partial_t f + \mathbf{v} \cdot \nabla_r f = -\nu(f - f^{LM}), \tag{2.69}$$

where ν denotes the mean collision frequency, independent of the microscopic velocity $\mathbf{v}$. The BGK collision operator in Eq. (2.69) is a nonlinear operator: the nonlinearity occurs because the local Maxwellian f^{LM} is parameterized by the values of the fields ρ, $\mathbf{u}$, T, which are obtained by integrating the distribution function itself, as also shown in Eqs. (2.52). In the applications, the linearized version of the BGK operator is mostly used. Considering small deviations from global equilibrium characterized by the values $[\rho_0, \mathbf{u}_0 = 0, T_0]$, the linearized BGK collision operator attains the form

$$L_{\mathrm{BGK}} = -\nu \left[h - \langle \tilde{\psi} | h \rangle \tilde{\psi}(\mathbf{c}) \right], \tag{2.70}$$

where $\mathbf{c} = \mathbf{v}/v_T$ is the dimensionless peculiar velocity and $v_T = \sqrt{2k_B T_0/m}$ is the equilibrium thermal velocity. The functions $\tilde{\psi}(\mathbf{c}) = [\tilde{\psi}_1, \tilde{\psi}_2, \tilde{\psi}_3]$ read

$$\tilde{\psi}(\mathbf{c}) = \left[1, \sqrt{2}\mathbf{c}, \sqrt{\frac{2}{3}} \left(c^2 - \frac{3}{2} \right) \right], \tag{2.71}$$

and they are mutually orthonormal with respect to the scalar product defined in Eq. (2.65). Equation (2.70) can also be formally written in the form

$$L_{\mathrm{BGK}} = -\nu \left[1 - P_0 \right] h, \tag{2.72}$$

where P_0 denotes the projection operator onto the subspace of $\mathscr{H}$ spanned by the functions $\tilde{\psi}$.

References

1. R. Balescu, *Equilibrium and nonequilibrium statistical mechanics* (Wiley, 1975).
2. L. E. Reichl, *A modern course in statistical physics* (University of Texas Press, Austin, 1980).
3. C. Cercignani, R. Illner and M. Pulvirenti, *The Mathematical Theory of Dilute Gases*, (Springer-Verlag, Berlin, 1994).
4. G. F. Mazenko, *Nonequilibrium Statistical Mechanics* (Wiley-VCH, 2006).
5. J.-P. Hansen and I.R. McDonald, *Theory of Simple Liquids* (Academic Press, 2006).
6. C. Cercignani, *Theory and Application of the Boltzmann Equation* (Scottish Academic Press, Edinburgh, 1975).
7. S. Chapman and T. G. Cowling, *The Mathematical Theory of Nonuniform Gases*, (Cambridge University Press, New York, 1970).
8. M. Bixon and R. Zwanzig, Boltzmann–Langevin Equation and Hydrodynamic fluctuations, *Phys. Rev.* **187**, 1 (1969).
9. H. Ueyama, The stochastic Boltzmann equation and hydrodynamic fluctuations, *J. Stat. Phys.* **22**, 1 (1980).
10. R. F. Fox and G. E. Uhlenbeck, Fluctuation Theory for the Boltzmann Equation, *Phys. Fluids* **13**, 2881 (1970).
11. L. D. Landau and E. M. Lifshitz, *Fluid Mechanics* (Pergamon Press, 1959).
12. H. C. Öttinger, *Betond Equilibrium Thermodynamics* (Wiley, 2005).

13. P.L. Bhatnagar, E.P. Gross and M. Krook, A Model for Collision Processes in Gases. I. Small Amplitude Processes in Charged and Neutral One-Component Systems, *Phys. Rev.* **94**, 511 (1954).

14. I.V. Karlin, M. Colangeli and M. Kröger, Exact linear hydrodynamics from the Boltzmann Equation, *Phys. Rev. Lett.* **100**, 214503 (2008).

15. A. Bellouquid, Global existence and large-times behavior for BGK model for a gas with nonconstant cross section, *Transp. Theory Stat. Phys.* **32**, 157 (2003).

Chapter 3
Methods of Reduced Description

In this chapter, we will review some analytical methods that make it possible to determine approximate solutions of the Boltzmann equation. In particular, we will discuss the structure of the Hilbert and Chapman–Enskog perturbation techniques and will also outline the essential features of the invariant manifold method, which stems from the assumption of time scale separation and, unlike the former methods, is also applicable beyond the strict hydrodynamic limit. Before reviewing the wealth of different techniques, it is worth investigating in greater depth the role of the different time scales in a particle system, which is one of the main ingredients underlying the onset of collective behavior.

3.1 The Bogoliubov Hypothesis and Macroscopic Equations

Let $\mathbf{M}_f = [M_1, \mathbf{M}_2, M_3]$ denote the lower-order moments of the single-particle distribution function $f(\mathbf{r}, \mathbf{v}, t)$ (2.24), defined as

$$\mathbf{M}_f(\mathbf{r}, t) = \int \psi(\mathbf{v}) f(\mathbf{r}, \mathbf{v}, t) \mathrm{d}\mathbf{v}, \tag{3.1}$$

where $\psi(\mathbf{v})$ are the collisional invariants (2.43). Thus, we have

$$M_1 = \rho(\mathbf{r}, t) = \int f(\mathbf{r}, \mathbf{v}, t) \mathrm{d}\mathbf{v},$$

$$\mathbf{M}_2 = \rho(\mathbf{r}, t) \mathbf{u}(\mathbf{r}, t) = \int m\mathbf{v} f(\mathbf{r}, \mathbf{v}, t) \mathrm{d}\mathbf{v},$$

$$M_3 = \frac{1}{2}\rho(\mathbf{r}, t) u^2 + \rho(\mathbf{r}, t) e(\mathbf{r}, t) = \int \frac{m}{2} v^2 f(\mathbf{r}, \mathbf{v}, t) \mathrm{d}\mathbf{v}, \tag{3.2}$$

where $\rho(\mathbf{r}, t) = mn(\mathbf{r}, t)$ is the mass density. The first of Eq. (3.2) indicates in particular that the number density $n(\mathbf{r}, t)$, besides corresponding to the phase space

M. Colangeli, *From Kinetic Models to Hydrodynamics*, SpringerBriefs in Mathematics,   23
DOI: 10.1007/978-1-4614-6306-1_3, © Matteo Colangeli 2013

average of the microscopic one-particle density (2.15), turns out also to be given by the lowest-order moment of the single-particle distribution $f(\mathbf{r}, \mathbf{v}, t)$ obtained from Eqs. (2.14) and (2.24). Moreover, in (3.2), we introduced the quantity

$$e(\mathbf{r}, t) = \frac{3}{2} \frac{k_B}{m} T(\mathbf{r}, t),$$

which denotes the internal energy per unit mass. The hydrodynamic fields $[\rho, \mathbf{u}, T]$, which correspond to mass density, bulk velocity and temperature, are related to the moments $\mathbf{M}_f$ by

$$\rho = M_1, \quad \mathbf{u} = M_1^{-1}\mathbf{M}_2, \quad T = \frac{2m}{3k_B M_1}(M_3 - \frac{1}{2}M_1^{-1}\mathbf{M}_2 \cdot \mathbf{M}_2). \tag{3.3}$$

We focus in this work on a special class of distribution functions called *normal solutions* of the Boltzmann equation [1]. These are distribution functions whose dependence on the variables $(\mathbf{r}, t)$ is parameterized by a set of fields $\mathbf{x}(\mathbf{r}, t)$, which typically correspond to the hydrodynamic fields themselves, but in certain cases may also include higher-order moments of the distribution function, as in Grad's moment method [2–4]. Let us therefore begin by writing the single-particle distribution function in the form

$$f(\mathbf{r}, \mathbf{v}, t) = f(\mathbf{x}(\mathbf{r}, t), \mathbf{v}). \tag{3.4}$$

A physical rationale behind Eq. (3.4) can be traced back to the Bogoliubov hypothesis [5–7], which assumes that three different time scales, labeled respectively τ_{int}, τ_{mf}, and τ_{macro}, characterize the relaxation of a gas toward equilibrium. The time interval τ_{int} is the time during which two molecules are in each other's interaction domain; τ_{mf} denotes the mean time between collisions, and τ_{macro} corresponds to the average time needed for a molecule to traverse the container in which the gas is confined. The Bogoliubov hypothesis states that the time scale separation

$$\tau_{int} \ll \tau_{mf} \ll \tau_{macro} \tag{3.5}$$

is a prerequisite for the onset of hydrodynamic behavior in a gas. In a similar manner, it is possible to introduce three corresponding displacements, denoted by λ_{int}, λ_{mf}, λ_{macro}. The displacement λ_{int} corresponds to the range of particle interaction, λ_{mf} is the mean free path, and λ_{macro} denotes a macroscopic reference length, such as the edge length of the confining container. Typical values of these parameters are listed in Table 3.1.

Using the notation introduced in Sect. 2.2, we denote by $\mathbf{z}$ the pair $(\mathbf{r}, \mathbf{v})$. The Bogoliubov hypothesis therefore concerns the functional dependence of the phase space density $F(\mathbf{z}_1, \ldots, \mathbf{z}_N, t)$ relevant to the three stages in the process of relaxation of the gas toward equilibrium; cf. also Table 3.2.

In the initial interval, there is no collisional exchange between the particles, and the gas experiences no equilibrating force. Bogoliubov conjectured that in such an

Table 3.1 Bogoliubov length and time intervals for a gas with mean molecular speed of 300 m/s at standard condition (container edge length $\lambda_{macro} = 3$ cm); cf. [5]

	λ_{int}	λ_{mf}	λ_{macro}
cm	3×10^{-8}	3×10^{-5}	3
s	10^{-12}	10^{-9}	10^{-4}

Table 3.2 Epochs in the Bogoliubov hypothesis

$t < \tau_{int}$	Initial stage
$\tau_{int} < t < \tau_{mf}$	Kinetic stage
$t \sim \tau_{macro}$	Hydrodynamic stage

In describing the particle system on time intervals $t < \tau_{int}$, the full phase space description is required. If the description is confined between τ_{int} and τ_{mf}, one can employ the framework of kinetic theory, and the Boltzmann equation provides an efficient statistical description of the dynamics of a sufficiently dilute particle system. Finally, for $t \simeq \tau_{macro}$, the dynamics of the distribution function is driven by the evolution of the hydrodynamic fields

initial stage, the full N-particle density is required to properly describe the state of the gas. In the kinetic stage, the molecules experience a sequence of collisions, which give rise to the onset of a local equilibrium in the gas. According to Bogoliubov's hypothesis, during this stage, all s-particle marginals introduced in Sect. 2.1 may be expressed as functionals of the single-particle density, i.e.,

$$F_s = F_s(\mathbf{z}_1, \ldots, \mathbf{z}_s, F(\mathbf{z}_1, t)), \tag{3.6}$$

where the time dependence of F_s is entirely contained in $F_1(\mathbf{z}_1, t)$. For instance, if the particles are statistically independent from each other, then F_s factorizes as follows:

$$F_s = \prod_{i=1}^{s} F_1(\mathbf{z}_i, t). \tag{3.7}$$

Finally, in the hydrodynamic stage, the relevant time scale is τ_{macro}, which characterizes the time evolution of the macroscopic variables $\mathbf{x}(\mathbf{r}, t)$. It is worth noticing that in the course of the relaxation, a crucial loss of information occurs [5]: while in the initial stage, the microscopic state of the particle system is described by the full phase space density, close to equilibrium, the statistical description is suitably afforded only in terms of the single-particle distribution function, parameterized by the variables $\mathbf{x}$.

We also remark that the condition $\tau_{int} \ll \tau_{mf}$ is essential for writing the Boltzmann equation in the form (2.40). As mentioned in Sect. 2.2, the distribution function $f(\mathbf{r}, \mathbf{v}, t)$ needs to be regarded as an *average* of the single-particle distribution in a time interval dt, with

$$\tau_{int} < dt < \tau_{mf}. \tag{3.8}$$

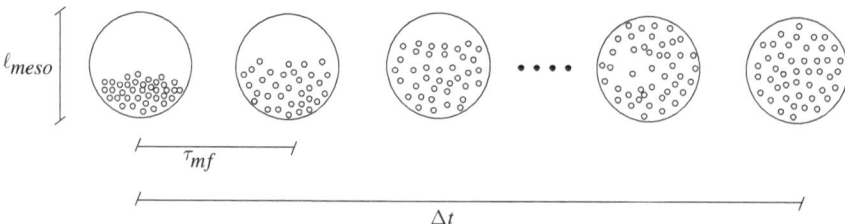

Fig. 3.1 The onset of local thermodynamic equilibrium in mesoscopic cells of size ℓ_{meso} taking place after a time interval Δt. Local equilibrium results from a large number of collisions occurring on time scales of order τ_{mf}. The hydrodynamic fields attain a local value within each of the cells and evolve on a time scale τ_{macro} (not shown in the picture) much larger than Δt, according to the time scale separation hypothesis.

This condition, in fact, allows us to disregard the variation experienced by the distribution function of the hitting particle, $f(\mathbf{r}, \mathbf{v}_1, t)$, during the time τ_{int} of interaction with the target particle. If the condition (3.8) does not hold, one should use in Eq. (2.40) the distribution function $f(\mathbf{r}, \mathbf{v}_1, t - \tau_{int})$ evaluated at an earlier time, and the rate of change of f at time t would depend not only on the instantaneous value of f, but also on its previous history [8]. This would make the Boltzmann equation a *non-Markovian* process. We see, then, that the separation between τ_{int} and τ_{mf} allows us to identify an intermediate scale dt, that guarantees the Markovian character of the Boltzmann equation. Moreover, in Sect. 2.3, we also introduced the time interval Δt, defined as the characteristic mesoscopic time scale characterizing the onset of local equilibrium. A macroscopic description of a particle system based on the hydrodynamic fields $\mathbf{x}(\mathbf{r}, t)$ can be obtained by confining the description to time scales not inferior to Δt, which is intermediate between τ_{mf} and the macroscopic scale τ_{macro}. The role of the time scale Δt can be better understood by introducing a partition of the volume of the gas into mesoscopic cells of linear size ℓ_{meso}; cf. Fig. 3.1. Local equilibrium is reached in the cells after the time interval Δt. Therefore, although the hydrodynamic variables may vary over macroscopic length and time scales, within each cell they obey, after the time interval Δt, the usual relations of equilibrium thermodynamics [9].

The dimensionless parameter ε, which we already encountered in Sect. 2.3, is the Knudsen number [1, 10] and is defined as the ratio of λ_{mf} to λ_{macro}. The aim of a generalized hydrodynamic theory is to extend the macroscopic description to finite Knudsen numbers, i.e., beyond the the hydrodynamic limit, corresponding to $\varepsilon \ll 1$. This, in turn, requires a proper estimate, for the model under consideration, of the magnitude of the length scale ℓ_{meso} below which the notion of local thermodynamic equilibrium, and consequently the hydrodynamic formalism, is lost. The Maxwell-molecules gas, in particular, is amenable to an investigation of the properties of the short-wavelength fluctuations of the hydrodynamic fields. The numerical analysis reported in Chap. 5 shows the presence of a critical length scale that marks the range of validity of the hydrodynamic description at short scales and is consistent with the discussed role of local equilibrium in the onset of collective behavior in a gas.

Thus, if we consider macroscopic length and time scales compatible, respectively, with ℓ_{meso} and Δt, it makes sense to discuss the derivation of macroscopic equations from the Boltzmann equation and to investigate their properties. To this end, we integrate Eq. (2.40), multiplied by the collision invariants (2.43), over the velocity space, and obtain

$$\partial_t \rho = -\nabla_{\mathbf{r}}(\rho \mathbf{u}),$$
$$\partial_t (\rho \mathbf{u}) = \nabla_{\mathbf{r}} \cdot (\rho \mathbf{u}\mathbf{u} + \mathbf{P}),$$
$$\partial_t \left[\rho \left(\frac{1}{2} u^2 + e \right) \right] = -\nabla_{\mathbf{r}} \cdot \left[\rho \mathbf{u} \left(\frac{1}{2} u^2 + e \right) + \mathbf{P} \cdot \mathbf{u} + \mathbf{q} \right], \qquad (3.9)$$

where the nonhydrodynamic fields $\mathbf{P}$ and $\mathbf{q}$, called the pressure tensor and heat flux, are defined as

$$\mathbf{P} = \int m(\mathbf{v} - \mathbf{u})(\mathbf{v} - \mathbf{u}) f(\mathbf{r}, \mathbf{v}, t) \mathrm{d}\mathbf{v}, \qquad (3.10)$$

$$\mathbf{q} = \int m(\mathbf{v} - \mathbf{u})(v^2 - 2\mathbf{v} \cdot \mathbf{u} + u^2) f(\mathbf{r}, \mathbf{v}, t) \mathrm{d}\mathbf{v}. \qquad (3.11)$$

The pressure tensor can be written as $\mathbf{P} = p\mathbf{I} + \boldsymbol{\sigma}$, where $\mathbf{I}$ is the identity matrix; the scalar p, defined as

$$p = \frac{1}{3} tr\, [\mathbf{P}] = nk_B T = \frac{2}{3}\rho e, \qquad (3.12)$$

corresponds to the hydrostatic pressure; and $\boldsymbol{\sigma}$ (not to be confused with the differential scattering cross section introduced in Chap. 2) is a symmetric tensor (it is also typically a traceless one, depending on the magnitude of the bulk viscosity, defined below). Recalling the splitting, introduced in Eq. (2.63), of the distribution function f into its equilibrium and nonequilibrium components, it is possible to show that in general, p carries the contribution of the only *local* Maxwellian component, whereas $\boldsymbol{\sigma}$ depends on the nonequilibrium contribution to the single-particle distribution function. Analogously, the heat flux vector $\mathbf{q}$ results only from the nonequilibrium component of the single-particle distribution function. A visible feature of Eq. (3.9) is that the equations are not closed, because of the presence of the nonhydrodynamic fields $\boldsymbol{\sigma}$ and $\mathbf{q}$. As discussed by C. Cercignani in [1], Eq. (3.9) "constitute an empty scheme, since there are 5 equations for 13 quantities. In order to have useful equations, one must have some expressions for $\boldsymbol{\sigma}$ and $\mathbf{q}$ in terms of ρ, $\mathbf{u}$ and e. Otherwise, one has to go back to the Boltzmann equation (2.40) and solve it; and once it has been done, everything is done, and Eq. (3.9) are useless!" This corresponds to the well-known problem of seeking a suitable closure to the macroscopic equations. This problem can be tackled either from the kinetic theory standpoint, i.e., by employing some model reduction or coarse-graining techniques [3], or from a purely macroscopic perspective, i.e., by employing macroscopic balance or phenomenological relations that disregard the underlying particle-like picture. In particular, the following set of constitutive equations written in component notation,

$$\sigma_{i,j} = 0, \tag{3.13}$$

$$q_i = 0, \tag{3.14}$$

with $i, j = 1, 2, 3$, yields the so-called Euler equations of inviscid hydrodynamics, which can also be derived from the Boltzmann equation by retaining only the Maxwellian contribution to the distribution function. The Navier-Stokes–Fourier (NSF) equations are instead obtained from the following constitutive equations:

$$\sigma_{i,j} = -\eta \left(\frac{\partial u_i}{\partial r_j} + \frac{\partial u_j}{\partial r_i} \right) + \left(\frac{2}{3}\eta - \zeta \right) \frac{\partial u_k}{\partial r_k} \delta_{ij}, \tag{3.15}$$

$$q_i = -\lambda \frac{\partial T}{\partial r_i}, \tag{3.16}$$

where we used the repeated index summation convention, and η, ζ, λ correspond to the transport coefficients called respectively shear viscosity, bulk viscosity (usually negligible), and thermal conductivity. The NSF equations deserve a special mention in fluid dynamics, because Eqs. (3.15) and (3.16) may be derived not only from the macroscopic principles of conservation of mass, momentum, and energy, but also, rigorously, from kinetic theory [1, 10, 11]. The latter derivation can be performed using some perturbative schemes, such as those discussed in the next section, which refer to the hydrodynamic limit of the Boltzmann equation.

3.2 The Hilbert and the Chapman–Enskog Methods

We provide here an overview of the Hilbert and Chapman–Enskog (CE) methods of solution of the Boltzmann equation in the hydrodynamic limit. The reader is referred to [1, 12] for an exhaustive treatment of the subject. To simplify the notation, we omit hereinafter in this section the hat over the single-particle distribution referring to a rescaled Boltzmann equation of the form (2.59), introduced in Sect. 2.3.
In the Hilbert method, the normal solutions are expanded in powers of the Knudsen number ε, i.e.,

$$f = \sum_{i=0}^{\infty} \varepsilon^i f^{(i)}, \tag{3.17}$$

which, substituted in Eq. (2.59), yields a sequence of integral equations

$$Q\left(f^{(0)}, f^{(0)}\right) = 0, \tag{3.18}$$

$$L f^{(1)} = (\partial_t + \mathbf{v} \cdot \nabla_{\mathbf{r}}) f^{(0)}, \tag{3.19}$$

$$L f^{(2)} = (\partial_t + \mathbf{v} \cdot \nabla_{\mathbf{r}}) f^{(1)} - 2Q\left(f^{(0)}, f^{(1)}\right), \tag{3.20}$$

to be solved order by order. Here L denotes the linearized collision integral defined in Eq. (2.64). From Eq. (3.18), it follows that $f^{(0)}$ corresponds to the local Maxwellian (2.54). The Fredholm alternative applied to (3.19) results in the following [12]:

- Solvability condition,

$$\int (\partial_t + \mathbf{v} \cdot \nabla_{\mathbf{r}}) f^{(0)} \psi(\mathbf{v}) d\mathbf{v} = 0, \tag{3.21}$$

which corresponds to the Euler equations described by Eqs. (3.13) and (3.14).
- General solution $f^{(1)} = f^{(1),1} + f^{(1),2}$, where $f^{(1),1}$ denotes the special solution to the linear integral equation (3.19), and $f^{(1),2}$ is a yet undetermined linear combination of the summational invariants.
- The solvability condition, when applied to Eq. (3.20), yields $f^{(1),2}$, which is obtained from solving the linear hyperbolic differential equations

$$\int (\partial_t + \mathbf{v} \cdot \nabla_{\mathbf{r}}) \left(f^{(1),1} + f^{(1),2}\right) \psi(\mathbf{v}) d\mathbf{v} = 0. \tag{3.22}$$

Hilbert showed that this procedure can be applied up to an arbitrary order n, so that the function $f^{(n)}$ is determined from the solvability condition applied at the $(n + 1)$th order [12]. Loosely speaking, the description provided by the Hilbert method is essentially in terms of the Euler equations, but it is supplemented by corrections that can by computed by solving linearized equations [1]. It is also worth remarking that the Hilbert method cannot provide uniformly valid solutions, which it can be understood by noticing the singular manner in which the Knudsen number enters the rescaled Boltzmann equation (2.59). Nevertheless, a truncated Hilbert expansion can reproduce, with arbitrary accuracy, the solution of the Boltzmann equation in a properly chosen region of time–space, provided that ε is sufficiently small.

The CE approach, developed by D. Enskog and S. Chapman, is based instead on an expansion of the time derivatives of the hydrodynamic variables, rather than seeking the time–space dependence of these functions, as in the Hilbert method. Also the CE method starts with the singularly perturbed Boltzmann equation (2.59), and with the expansion (3.17). Nevertheless, the procedure of evaluation of the functions $f^{(i)}$ is different, and reads as follows:

$$Q\left(f^{(0)}, f^{(0)}\right) = 0, \tag{3.23}$$

$$L f^{(1)} = -Q\left(f^{(0)}, f^{(0)}\right) + \left(\partial_t^{(0)} + \mathbf{v} \cdot \nabla_{\mathbf{r}}\right) f^{(0)}. \tag{3.24}$$

Equation (3.23) implies, as in the Hilbert method, that the function $f^{(0)}$ is the local Maxwellian. The operator $\partial_t^{(0)}$ is defined from the expansion of the right-hand side of the hydrodynamic equations,

$$\partial_t^{(0)} \mathbf{M}_f = - \int \psi(\mathbf{v}) \mathbf{v} \cdot \nabla_\mathbf{r} f^{(0)} d\mathbf{v}. \tag{3.25}$$

Equation (3.25) corresponds to the inviscid Euler equations, and $\partial_t^{(0)}$ acts on various functions $g(\rho, \rho\mathbf{u}, e)$ according to the chain rule

$$\partial_t^{(0)} g = \frac{\partial g}{\partial \rho} \partial_t^{(0)} \rho + \frac{\partial g}{\partial(\rho\mathbf{u})} \partial_t^{(0)} \rho\mathbf{u} + \frac{\partial g}{\partial e} \partial_t^{(0)} e, \tag{3.26}$$

whereas the time derivatives $\partial_t^{(0)}$ of the hydrodynamic fields are expressed using the right-hand side of (3.25). Finally, the method requires that the hydrodynamic variables obtained by integrating over the velocity space the function $f^{(0)} + \varepsilon f^{(1)}$ coincide with the parameters of the local Maxwellian $f^{(0)}$:

$$\int \psi(\mathbf{v}) f^{(1)} d\mathbf{v} = 0. \tag{3.27}$$

Thus, one finds that the first correction, $f^{(1)}$, adds the terms

$$\partial_t^{(1)} \mathbf{M}_f = - \int \psi(\mathbf{v}) \mathbf{v} \cdot \nabla_\mathbf{r} f^{(1)} d\mathbf{v} \tag{3.28}$$

to the time derivatives of the hydrodynamic fields. These novel terms yield the dissipative NSF hydrodynamics. However, higher-order corrections of the CE method, which result in hydrodynamic equations with higher derivatives (the so-called Burnett and super-Burnett hydrodynamics), are affected by severe difficulties, mainly related to the onset of instabilities of the solutions [13–15].

3.3 Grad's Moment Method

An alternative technique to solving the Boltzmann equation was proposed by H. Grad, and is known as Grad's moment method [4]. The essence of the method relies on the time scale separation hypothesis, introduced in Sect. 3.1:

- During the fast evolution, which occurs on a time scale of the order of the mesoscopic time scale Δt, a set of distinguished moments $\mathbf{x}$, cf. Sect. 3.1, does not change significantly in comparison to the rest of the higher-order "fast" moments of f, denoted by $\mathbf{y}$.

- Toward the end of the fast evolution, the values of the moments **y** become determined by the values of the distinguished moments **x**.
- During a time interval of order τ_{macro}, the dynamics of the distribution function is governed by the dynamics of the distinguished moments, while the rest of moments remain to be determined by the distinguished moments [12].

In Grad's moment method, the distribution function is expanded as

$$f(\mathbf{x}, \mathbf{v}) = f^{\text{LM}}(\rho, \mathbf{u}, e, \mathbf{v}) \left[1 + \sum_{k=1}^{N} a_k(\mathbf{x}) H_k(\mathbf{v} - \mathbf{u}) \right], \tag{3.29}$$

where $H_k(\mathbf{v} - \mathbf{u})$ are Hermite tensor polynomials, orthogonal with respect to a weight given by the Maxwellian distribution f^{LM}, whereas the coefficients a_k are known functions of the distinguished moments **x**. The fast moments **y** are assumed to be functions of **x**, i.e., $\mathbf{y} = \mathbf{y}(\mathbf{x})$. By inserting Eq. (3.29) into the Boltzmann equation (2.40) and using the orthogonality of the Hermite polynomials with respect to the Maxwellian distribution f^{LM}, one can determine the time evolution of the set of distinguished moments **x**. According to Grad's argument, this approximation can be refined by extending the set of distinguished moments **x**. The best known approximation is perhaps Grad's 13-moment approximation, which will be studied in Chap. 6. It consists of a set of time evolution equations for the standard five hydrodynamic ones, the five components of the symmetric traceless stress tensor σ, and the three components of the heat flux vector **q**. It is also worth mentioning that the pioneering contributions of H. Grad to kinetic theory paved the way for the development of the theory of extended irreversible thermodynamics [2, 3, 12].

3.4 The Invariant Manifold Theory

The invariant manifold method can be considered a generalization of the theory of normal solutions, which is inherent in the Hilbert and CE expansions [12]. The method is based on a projector operator formalism [16–18] that confines the phase space dynamics onto a manifold of slow motion and disregards the fluctuations of the fast variables. The same approach is also at the basis of Haken's slaving principle and the procedure of adiabatic elimination of fast variables in stochastic processes [19, 20]. We will restrict, hereinafter, the description to the space of single-particle distribution functions. The time evolution of the system is assumed to resemble the picture given in Sect. 3.3: from the initial condition, the system quickly approaches a small neighborhood of the invariant manifold, and from then onward, it proceeds slowly along such a manifold with a characteristic time scale of order τ_{macro}. The main geometric structures that characterize the invariant manifold theory are illustrated in Fig. 3.2.

We summarize in the sequel the essential mathematical framework that will be used in the next chapters. Let U be the phase space, and $\Omega \subset U$ an ansatz manifold that corresponds to the current approximation to the invariant manifold to be sought.

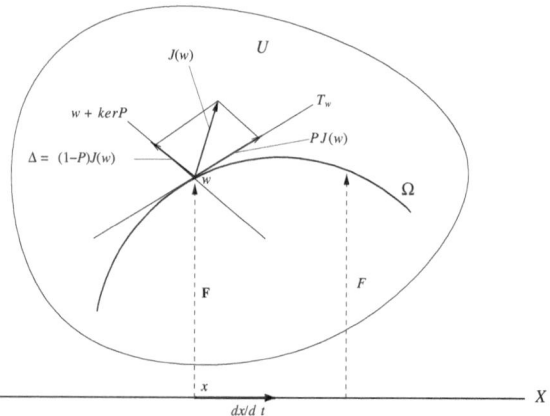

Fig. 3.2 The geometric structures of the invariant manifold method: U is the space of distribution functions, $J(w)$ is the vector field of the system under consideration, Ω is an ansatz manifold, X is the space of macroscopic variables (coordinates on the manifold), the map F maps points $\mathbf{x} \in X$ into the corresponding points $w = F(\mathbf{x})$, T_w is the tangent space to the manifold Ω at the point w, $PJ(w)$ is the projection of the vector $J(w)$ onto the tangent space T_w, $d\mathbf{x}/dt$ describes the induced dynamics on X, Δ is the defect of invariance, and the affine subspace $w + \ker P$ is the plane of fast motions [12]

We denote by $J(w)$ the vector field

$$\frac{dw}{dt} = J(w), \quad w \in U, \tag{3.30}$$

which generates the dynamics in U. Let X be the linear space of the macroscopic variables $\mathbf{x}$, which act as coordinates on the manifold Ω, described as the image of the map $F : X \to U$. We remark that the choice of the space X of macroscopic variables is a crucial step of the method: the corrections of the current ansatz manifold correspond to the images of various maps F for a given X. Let us also denote by

$$D_{\mathbf{x}} F = \left.\frac{\partial w}{\partial \mathbf{x}}\right|_{w=F(\mathbf{x})} \tag{3.31}$$

the derivative of the map F with respect to the set of distinguished variables. We indicate by $\Pi : U \to X$ a regular map that satisfies the condition

$$\Pi \circ F = 1, \tag{3.32}$$

with 1 the identity operator, and by $D_w \Pi$ the functional derivative of the map $\mathbf{x} = \Pi(w)$ computed at the point w. Thus, the time evolution of the distinguished variables $\mathbf{x}$ reads

$$\frac{d\mathbf{x}}{dt} = D_w \Pi J(F(\mathbf{x})), \tag{3.33}$$

where $d\mathbf{x}/dt$ is an element of the tangent space to X. Therefore, joining Eqs. (3.31) and (3.33), one obtains

$$\frac{dw}{dt}\Big|_{w=F(\mathbf{x})} = D_\mathbf{x}F \cdot \frac{d\mathbf{x}}{dt} = [D_\mathbf{x}F \circ D_w\Pi] \, J(F(\mathbf{x})) = PJ\,(F(\mathbf{x})), \qquad (3.34)$$

where the operator

$$P = D_\mathbf{x}F \circ D_w\Pi$$

projects $J(f)$ onto T_w, which denotes the tangent space to the manifold Ω at the point w. In particular, the projector P determines a decomposition of the motion near Ω: $w + \ker[P]$ is the plane of fast motion and T_w the plane of slow motion. We use the term *slow invariant manifolds* to describe those maps F that satisfy the condition (3.32) and solve the *invariance equation*

$$\Delta(F) = (1 - P)J(F) = 0, \qquad (3.35)$$

which is a differential equation for the unknown map F. The solutions of Eq. (3.35) are "invariant" in the sense that the vector field $J(F)$ is tangent to the manifold $\Omega = F(X)$ for each point $w \in \Omega$. A crucial aspect of the method concerns the definition of the projector P. A. N. Gorban et al. introduced, in [12], the thermodynamic projector, which characterizes, in a thermodynamic sense, the plane of fast motion $w + \ker P$: the physical entropy grows during the fast motion, and the point w is the maximum point of entropy along the plane $w + \ker P$.

The geometric setting described above can be readily adapted to the Boltzmann equation theory. To this end, one identifies w with the single-particle distribution function f, $\mathbf{x} = \Pi(f)$ denotes a set of distinguished fields that parameterize f, and F becomes a "closure," i.e., a distribution function parameterized by the variables $\mathbf{x}$. Moreover, the vector field $J(f)$ attains the form

$$J(f) = -\mathbf{v} \cdot \nabla_\mathbf{r} f + Q(f, f),$$

whereas $D_f\Pi$ reads

$$D_f\Pi[\cdot] = \int \psi(\mathbf{v})[\cdot]d\mathbf{v},$$

with $\psi(\mathbf{v})$ defined in (2.44). Therefore, the thermodynamic projector P, which depends on f, attains in this case the structure

$$P[\cdot] = \frac{\partial f}{\partial \mathbf{x}} \cdot \int \psi(\mathbf{v})[\cdot]d\mathbf{v}. \qquad (3.36)$$

The invariance Eq. (3.35) constrains the kinetic evolution of the distribution function to coincide with its "macroscopic" evolution, ruled by the projector P (3.36) and driven by the dynamics of the distinguished variables **x**. It should be noticed that the method does not require the smallness of the parameter ε; hence it is not restricted to the strict hydrodynamic limit. In particular, as discussed in Chap. 6, it is possible to prove, for some models, that the solutions of (3.35) correspond to an exact summation of the corresponding CE expansion, hence provide a generalized hydrodynamic description valid also at finite Knudsen numbers [12–14, 21, 22]. A straightforward application of the formalism described above is obtained by considering an ansatz manifold Ω_{LM} given by the locally five-dimensional manifold of local Maxwellians (2.54). We therefore take the set $\mathbf{M}_f$ of moments (3.2) as the coordinates **x** on this manifold. The manifold Ω_{LM} is commonly referred to as the *quasiequilibrium* manifold for the set of moments **x**, because f^{LM} corresponds to the unique solution of the variational problem

$$H(f) \to \min,$$

with $H(f)$ given in (2.48). We define the projector $P_{f^{LM}}$ onto the tangent space $T_{f^{LM}}$ as

$$P_{f^{LM}} J\left(f^{LM}\right) = \frac{\partial f^{LM}}{\partial \mathbf{M}_f} \cdot \int \psi(\mathbf{v}) J\left(f^{LM}\right) d\mathbf{v}. \tag{3.37}$$

Returning to the hydrodynamic variables $[\rho, \mathbf{u}, T]$ via the transformations (3.3), one obtains [12]

$$\Delta\left(f^{LM}(\rho, \mathbf{u}, T)\right) = f^{LM}(\rho, \mathbf{u}, T)\left[\left(\frac{m(\mathbf{v} - \mathbf{u})^2}{2k_B T} - \frac{5}{2}\right)(\mathbf{v} - \mathbf{u})\nabla_{\mathbf{r}}(\ln T) + \right.$$

$$\left. + \frac{m}{k_B T}\left((\mathbf{v} - \mathbf{u})(\mathbf{v} - \mathbf{u}) - \frac{1}{3}(\mathbf{v} - \mathbf{u})^2 \mathbf{I}\right)\nabla_{\mathbf{r}}\mathbf{u}\right]. \tag{3.38}$$

Equation (3.38) reveals that the quasiequilibrium manifold is not an invariant manifold of the Boltzmann equation, because temperature and bulk velocity gradients drive the invariant manifold away from local equilibrium. We remark that the particle system actually never reaches local equilibrium, or if it accidentally starts in that state, it moves away from it, due to the flow term $\left(\mathbf{v} \cdot \frac{\partial f}{\partial \mathbf{r}}\right)$ in the Boltzmann equation, whose effect is to smooth out the spatial inhomogeneities [8]. Yet for small Knudsen numbers, the flow term acts on time scales much larger than the collisions. Consequently, at all times, the instantaneous single-particle distribution function is very close to the local equilibrium one, given in (2.54). The latter may hence be regarded as a reference distribution in perturbation theories, such as those described above, that address the hydrodynamic limit of the Boltzmann equation.

References

1. C. Cercignani, *Theory and Application of the Boltzmann Equation* (Scottish Academic Press, Edinburgh, 1975).
2. D. Jou, J. Casas-Vázquez and G. Lebon, *Extended Irreversible Thermodynamics* (Springer, Berlin, 2010).
3. H. C. Öttinger, *Beyond Equilibrium Thermodynamics* (Wiley, New York, 2005).
4. H. Grad, On the kinetic theory of rarefied gases, *Comm. Pure and Appl. Math.* **2**, 331 (1949).
5. R. Liboff, *Kinetic Theory Classical, Quantum and Relativistic Descriptions* (Springer, New York, 2003).
6. N. N. Bogoliubov, in *Studies in Statistical Mechanics*, vol. 1, J. de Boer and G. E. Uhlenbeck (Wiley, New York, 1962).
7. R. Liboff, Generalized Bogoliubov hypothesis for dense fluids, *Phys. Rev. A* **31**, 1883 (1985).
8. R. Balescu, *Equilibrium and nonequilibrium statistical mechanics* (Wiley, New York, 1975).
9. D. Forster, *Hydrodynamic fluctuations, Broken Symmetry, and Corrrelation Functions* (W. A. Benjamin, New York, 1975).
10. S. Chapman and T. G. Cowling, *The Mathematical Theory of Nonuniform Gases*, (Cambridge University Press, New York, 1970).
11. M. Lachowicz, From kinetic to Navier-Stokes-type equations, *Appl. Math. Lett.* **10**, 19 (1997).
12. A.N. Gorban and I.V. Karlin, *Invariant Manifolds for Physical and Chemical Kinetics*, Lect. Notes Phys. **660** (Springer, Berlin, 2005).
13. M. Colangeli, I.V. Karlin and M. Kröger, From hyperbolic regularization to exact hydrodynamics for linearized Grad's equations, *Phys. Rev. E* **75**, 051204 (2007).
14. M. Colangeli, I.V. Karlin and M. Kröger, Hyperbolicity of exact hydrodynamics for three-dimensional linearized Grad's equations, *Phys. Rev. E* **76**, 022201 (2007).
15. A.V. Bobylev, Sov. Phys. Dokl. **27**, 29 (1982).
16. R. Zwanzig, Memory Effects in Irreversible Thermodynamics, *Phys. Rev.* **124**, 983 (1961).
17. D. N. Zubarev and V. P. Kalashnikov, Extremal properties of the nonequilibrium statistical operator, *Theor. Math. Phys.* **1**, 108 (1969).
18. H. Grabert, *Projection Operator Techniques in Nonequilibrium statistical Mechanics* (Springer, Berlin, 1982).
19. H. Haken, *Synergetics: an introduction: nonequilibrium phase transitions and self-organization in physics, chemistry and biology* (Springer, Berlin, 1978).
20. C. W. Gardiner, *Handbook of stochastic methods* (Springer, Berlin, 2004).
21. I.V. Karlin, Exact summation of the Chapman-Enskog expansion from moment equations, *J. Phys. A: Math. Gen.* **33**, 8037 (2000).
22. I.V. Karlin and A.N. Gorban, Hydrodynamics from Grad's equations: What can we learn from exact solutions?, *Ann. Phys. (Leipzig)* **11**, 783 (2002).

Chapter 4
Hydrodynamic Spectrum of Simple Fluids

In this chapter, we will focus on the statistical properties of a fluid from a macroscopic perspective. To this end, we will discuss the properties of the linearized version of the NSF equations of hydrodynamics and will introduce the correlation function formalism, which allows us to characterize the spectrum of fluctuations of the hydrodynamic variables. In particular, we will also clarify the structure of an experimentally accessible quantity, the dynamical structure factor, that is related to the intensity of the scattered radiation in both the hydrodynamic and free-particle limits. To fix the ideas, let us consider a fluid at rest, in thermodynamic equilibrium, that is hit by a mono-energetic beam of radiation, e.g., thermal neutrons. The intensity of the scattered radiation turns out to be a function, characteristic of the fluid, that depends on the angle of scattering and on the frequency of the outgoing beam. The scattering is caused by the presence of spontaneous microscopic fluctuations, induced by thermal excitations [1, 2], which can be described in terms of suitable correlation functions. Interestingly, the same functions also make it possible to capture the mathematical essence of the response of the system to a weak external field [3–5]. The reason for this connection stems from the fact that the dynamical processes that govern the spatial and temporal decay of the fluctuations also determine the linear response of the system. In particular, close to equilibrium and under the assumption of time reversibility, the Onsager–Machlup theory [6, 7] guarantees that a spontaneous fluctuation most likely follows a path that is the time reversal of the relaxation path induced by an external perturbation. The connection between spontaneous fluctuations and linear response to an external perturbation explains the relevance of the correlation function formalism in the study of transport phenomena in fluids and in unveiling the molecular origin of transport coefficients, even at high frequencies and large wave vectors [1]. The mathematical framework discussed in this chapter sets the stage for the development of a generalized hydrodynamic theory that we will pursue in the following chapters.

M. Colangeli, *From Kinetic Models to Hydrodynamics*, SpringerBriefs in Mathematics,   37
DOI: 10.1007/978-1-4614-6306-1_4, © Matteo Colangeli 2013

4.1 Correlations in Space and Time

Let us consider two microscopic dynamical variables $a(\mathbf{Q}, \mathbf{P}, \mathbf{r}_1, t_1)$ and $b(\mathbf{Q}, \mathbf{P}, \mathbf{r}_2, t_2)$, of the sort introduced in Sect. 2.1. We define the two-time correlation function $C_{a,b}$ between the phase functions a and b as

$$C_{a,b}(\mathbf{r}_1, \mathbf{r}_2, t_1, t_2) = \langle \delta a(\mathbf{Q}, \mathbf{P}, \mathbf{r}_1, t_1) \delta b(\mathbf{Q}, \mathbf{P}, \mathbf{r}_2, t_2) \rangle, \qquad (4.1)$$

where the average is taken with respect to a given density $F(\mathbf{Q}, \mathbf{P})$; cf. Eq. (2.6). In Eq. (4.1), we denoted by

$$\delta a(\mathbf{Q}, \mathbf{P}, \mathbf{r}_1, t_1) = a(\mathbf{Q}, \mathbf{P}, \mathbf{r}_1, t_1) - A(\mathbf{r}_1, t_1)$$

the fluctuation of the dynamical variable a with respect to its ensemble average $A \equiv \langle a \rangle$ (the same notation also applies to the variable b). The fluctuations can be regarded as spontaneous random departures from the average. We also note in passing that this standpoint is consistent with the deterministic evolution dictated by Hamiltonian dynamics if one regards the stochastic character of the fluctuations as induced by the unknown distribution of δa at time $t = 0$ [8]. As mentioned in Sect. 2.1, most observables of interest are obtained by integrating the empirical distribution function (2.13) over the momentum space:

$$a(\mathbf{Q}, \mathbf{P}, \mathbf{r}_1, t_1) = \int \psi_a(\boldsymbol{\xi}_1) \sum_{i=1}^{N} \delta\left(\mathbf{r}_1 - \mathbf{q}_i(t_1)\right) \delta\left(\boldsymbol{\xi}_1 - \mathbf{p}_i(t_1)\right) d\boldsymbol{\xi}_1, \qquad (4.2)$$

where $\psi_a(\boldsymbol{\xi}_1)$ denotes some known function of the momentum variable $\boldsymbol{\xi}_1$ (e.g., the elementary collision invariants, defined in Eq. (2.44)). From Eq. (4.2), we can rewrite Eq. (4.1) in the form

$$C_{a,b}(\mathbf{r}_1, \mathbf{r}_2, t_1, t_2) = \langle \psi_a(\boldsymbol{\xi}_1) \psi_b(\boldsymbol{\xi}_2) \rangle_{f_2} - \langle \psi_a(\boldsymbol{\xi}_1) \rangle_f \langle \psi_b(\boldsymbol{\xi}_2) \rangle_f, \qquad (4.3)$$

where $\langle \ldots \rangle_{f_2}$ denotes an integration over the momenta $\boldsymbol{\xi}_1$ and $\boldsymbol{\xi}_2$, taken with the distribution $f_2(\mathbf{r}_1, \boldsymbol{\xi}_1, t_1; \mathbf{r}_2, \boldsymbol{\xi}_2, t_2)$ given in Eq. (2.19). Instead, $\langle \ldots \rangle_f$ corresponds to an average over the single momentum space $\boldsymbol{\xi}_i$, with $i = 1, 2$, taken with the corresponding single-particle distribution $f(\mathbf{r}_i, \boldsymbol{\xi}_i, t_i)$. Let us introduce the Fourier transform of the fluctuation $\delta a(\mathbf{Q}, \mathbf{P}, \mathbf{r}, t)$:

$$a_{\mathbf{k}}(\mathbf{Q}, \mathbf{P}, t) = \int\limits_{-\infty}^{+\infty} \delta a(\mathbf{Q}, \mathbf{P}, \mathbf{r}, t) e^{-i\mathbf{k}\cdot\mathbf{r}} d\mathbf{r}. \qquad (4.4)$$

Correspondingly, the Fourier transform of Eq. (4.1) is given by [9]

$$C_{a,b}(\mathbf{k}_1, \mathbf{k}_2, t_1, t_2) = \left\langle a_{\mathbf{k}_1}(t_1) b_{-\mathbf{k}_2}(t_2) \right\rangle. \tag{4.5}$$

The structure of the correlation function $C_{a,b}$ considerably simplifies if the average in Eq. (4.1) is taken with respect to an equilibrium phase density $F_{eq}(\mathbf{Q}, \mathbf{P})$, such as the one in (2.10). In this case, the correlation function enjoys the properties of *stationarity* and of *translational invariance*, because it depends respectively only on the time difference $t = t_1 - t_2$ and on the spatial distance $\mathbf{r} = \mathbf{r}_1 - \mathbf{r}_2$. Thus, an equilibrium time correlation function obeys the relation

$$C_{a,b}(\mathbf{r}_1, \mathbf{r}_2, t_1, t_2) = C_{a,b}(\mathbf{r}, t), \tag{4.6}$$

which is mirrored, in Fourier space, by

$$C_{a,b}(\mathbf{k}_1, \mathbf{k}_2, t_1, t_2) = C_{a,b}(\mathbf{k}_1, t)\delta_{\mathbf{k}_1, \mathbf{k}_2}, \tag{4.7}$$

with $C_{a,b}(\mathbf{k}_1, t)$ denoting the spatial Fourier transform of $C_{a,b}(\mathbf{r}, t)$. Therefore, whenever any two-point function such as $C_{a,b}(\mathbf{r}_1, \mathbf{r}_2, t_1, t_2)$ is translation-invariant in real space, it corresponds to a two-point function that is diagonal in Fourier space, i.e., it is nonzero only when $\mathbf{k}_1 = \mathbf{k}_2$. Under the hypothesis of translational invariance, and denoting by $\mathbf{k}$ the wave vector, the further property of *isotropy* of the fluid entails that Eqs. (4.6) and (4.7) then depend only on $r \equiv |\mathbf{r}|$ and $k \equiv |\mathbf{k}|$. We will focus in the remainder of this chapter on the structure of equilibrium time correlation functions and will consider in particular the phase function given by the local number density $n(\mathbf{Q}, \mathbf{r})$, given in (2.15). For a uniform fluid, the van Hove function is defined as [9]

$$G(\mathbf{r}, t) = \left\langle \frac{1}{N} \sum_{i=1}^{N} \sum_{j=1}^{N} \delta\left(\mathbf{r} - \mathbf{r}_j(t) + \mathbf{r}_i(0)\right) \right\rangle. \tag{4.8}$$

This function can be endowed with a simple physical interpretation: $G(\mathbf{r}, t)d\mathbf{r}$ is the number of particles j in a region $d\mathbf{r}$ around $\mathbf{r}$ at time t, given that there was a particle i centered at the origin at time $t = 0$. The function $G(\mathbf{r}, t)$, which reduces at $t = 0$ to the *static correlation function*, can be split into two terms, referred to, respectively, as the *self* part and the *distinct* part:

$$G(\mathbf{r}, t) = G_s(\mathbf{r}, t) + G_d(\mathbf{r}, t), \tag{4.9}$$

with

$$G_s(\mathbf{r}, t) = \left\langle \frac{1}{N} \sum_{i=1}^{N} \delta\left(\mathbf{r} - \mathbf{r}_i(t) + \mathbf{r}_i(0)\right) \right\rangle, \tag{4.10}$$

$$G_d(\mathbf{r}, t) = \left\langle \frac{1}{N} \sum_{i=1}^{N} \sum_{j\neq1}^{N} \delta\left[\mathbf{r} - \mathbf{r}_j(t) + \mathbf{r}_i(0)\right] \right\rangle. \tag{4.11}$$

Hence $G_s(\mathbf{r}, 0) = \delta(\mathbf{r})$, and it can also be shown that $G_d(\mathbf{r}, 0) = (N/V)g(\mathbf{r})$, where, using the notation of Sect. 2.1, V is the volume of the system and $g(\mathbf{r})$ denotes the radial distribution function introduced after Eq. (2.17), whose spatial Fourier transform corresponds to the *static structure factor* $S_{n,n}(\mathbf{k})$. Let us introduce, then, the *intermediate scattering function*, defined as

$$F_{n,n}(\mathbf{k}, t) = \frac{1}{N}\langle n_{\mathbf{k}}(t)n_{-\mathbf{k}}\rangle, \tag{4.12}$$

which is related to $G(\mathbf{r}, t)$ by

$$F_{n,n}(\mathbf{k}, t) = \int_{-\infty}^{\infty} G(\mathbf{r}, t)e^{-i\mathbf{k}\cdot\mathbf{r}}d\mathbf{r}. \tag{4.13}$$

The time Fourier transform of the function (4.12) yields the *dynamic structure factor* (or power spectrum) $S_{n,n}(\mathbf{k}, \omega)$:

$$S_{n,n}(\mathbf{k}, \omega) = \int_{-\infty}^{\infty} F_{n,n}(\mathbf{k}, t)e^{-i\omega t}dt, \tag{4.14}$$

which is also related to the static structure factor $S_{n,n}(\mathbf{k})$ by the relation

$$S_{n,n}(\mathbf{k}) = F_{n,n}(\mathbf{k}, 0) = \int_{-\infty}^{\infty} S_{n,n}(\mathbf{k}, \omega)d\omega. \tag{4.15}$$

The function $S_{n,n}(\mathbf{k})$ can be directly measured from experiments of elastic scattering of neutrons or x-rays by a fluid. The experiment consists in irradiating a fluid with a beam of radiation with wave vector $\mathbf{k}_i$ and frequency ω. If the energy of the quanta is much larger than the characteristic excitation energy of the molecules, the scattering occurs without any change in the wave frequency, so that the magnitude of the wave vector is conserved (elastic scattering). We denote by $\mathbf{k} = \mathbf{k}_f - \mathbf{k}_i$ the momentum transfer; in elastic scattering experiments, the intensity of the scattered radiation, which is related to the differential scattering cross section, depends only on this variable and is proportional to $S_{n,n}(\mathbf{k})$ [8, 9]. On the other hand, experiments based on inelastic scattering involve as well a transfer of energy $\hbar\omega$, and it can be proved [9] that in this case, the differential scattering cross section at a given frequency ω is related to the structure factor $S_{n,n}(\mathbf{k}, \omega)$. Another relevant relation concerning $S_{n,n}(\mathbf{k}, \omega)$ is the *detailed balance* condition [1, 10]:

$$S_{n,n}(\mathbf{k}, \omega) = e^{-\beta\hbar\omega}S_{n,n}(-\mathbf{k}, -\omega), \tag{4.16}$$

which describes the thermal equilibrium between the radiation and the fluid. Equation (4.16) reveals in particular that the spectral distribution of a beam of radiation scattered by the fluid is *not* symmetric with respect to the energy gain and energy loss. Nevertheless, the symmetry is restored in the classical limit $\hbar \to 0$, in which $S_{n,n}(\mathbf{k}, \omega) = S_{n,n}(-\mathbf{k}, -\omega)$ holds [1]. Let us now discuss the form of the dynamic structure factor in the free-particle limit, which corresponds to $\varepsilon \gg 1$ and pertains to the regime, typical of an ideal gas, in which the particles move freely in the fluid at constant velocity. Because in this limit the positions of the particles are uncorrelated, the calculation of $S_{n,n}(\mathbf{k}, \omega)$ reduces to the calculation of the self part $G_s(\mathbf{r}, t)$ of the van Hove function (4.10). In particular, the probability that an ideal gas particle covers a distance $\mathbf{r}$ in a time t is given by the Maxwellian distribution that a particle has a velocity in the range $d\mathbf{u}$ around $\mathbf{u}$, where $\mathbf{u} = \mathbf{r}/t$. This leads to the expression

$$G_s(\mathbf{r}, t) = \frac{1}{\sqrt{\left(\pi v_T^2 t^2\right)^3}} e^{-\frac{r^2}{(v_T t)^2}}, \qquad (4.17)$$

where $v_T = \sqrt{2k_B T/m}$ is the thermal velocity introduced in Sect. 2.4. Therefore, from (4.17), one obtains

$$S_{n,n}(\mathbf{k}, \omega) = \frac{1}{\sqrt{\pi v_T^2 k^2}} e^{-\left(\frac{\omega}{v_T k}\right)^2}, \qquad (4.18)$$

which gives an asymptotic expression for the dynamical structure factor $S_{n,n}(\mathbf{k}, \omega)$ in the domain of large wave vectors and high frequencies. By contrast, the structure of $S_{n,n}(\mathbf{k}, \omega)$ pertaining to the opposite regime of small wave vectors and low frequencies, corresponding to the hydrodynamic limit, will be derived in Sect. 4.2. Before concluding this section, it is worth discussing briefly the relevant role of the correlation function formalism in the branch of statistical mechanics concerned with linear response theory [3]. This theory focuses on the statistical properties of many-particle systems weakly perturbed from equilibrium; it received a theoretical justification with the formulation of the *fluctuation–dissipation theorem* (FDT) [11, 12]. The literature reports, in fact, a variety of results referring to this theorem, which hold in rather different contexts, e.g., in deterministic Hamiltonian systems ("FDT of the first kind" [4, 12]), in stochastic dynamics ("FDT of the second kind"; see [13–15]), in deterministic dissipative systems ("generalized fluctuation–dissipation relations"; cf. [16]). We restrict ourselves here to the FDT of the first kind. We consider a system that at time $t = t_0$ is at equilibrium with a (time-independent) Hamiltonian $H(\mathbf{Q}, \mathbf{P})$. We also assume that the statistical properties of the system are described by the equilibrium density (2.10). Then at time $t = 0$, a perturbation $h(t)$ is switched on, and it modifies the Hamiltonian of the system into $H(\mathbf{Q}, \mathbf{P}) \to H(\mathbf{Q}, \mathbf{P}) - h(t)a(\mathbf{Q}, \mathbf{P})$, where $a(\mathbf{Q}, \mathbf{P})$ is a phase function conjugate to the external field $h(t)$. Correspondingly, the ensemble average of an observable $b(\mathbf{Q}, \mathbf{P})$ is transformed by the perturbation into $\langle b \rangle_{\mathrm{eq}} \to \langle b \rangle_h(t) = \langle b \rangle_{\mathrm{eq}} + \langle \Delta b \rangle_h(t)$, where the subscripts eq and h refer respectively to the averages taken with respect to the equilibrium density (2.10) and to the perturbed density induced by the external field. As long as the magnitude of the perturbation is small, one may expand $\langle b \rangle_h$ around $h = 0$ (i.e., around equilibrium) to obtain

$$\langle b \rangle_h(t) = \langle b \rangle_{\text{eq}} + \int_0^t \left. \frac{\delta \langle b \rangle_h(t)}{\delta h(s)} \right|_{h=0} h(s) ds = \langle b \rangle_{\text{eq}} + \int_0^t R_{b,a}(t-s)h(s) ds, \quad (4.19)$$

where $\delta \langle b \rangle_h(t)/\delta h(s)$ denotes the functional derivative of the ensemble average $\langle b \rangle_h$ with respect to the perturbation h, and $R_{b,a}(t-s)$ is the *response function*. The FDT of the first kind states that the response function $R_{b,a}(t-s)$, induced by a perturbation $h(s)$ occurring at the time $s \leq t$ is given by [3, 13]

$$R_{b,a}(t-s) = \beta \frac{\mathrm{d}}{\mathrm{d}s} \langle a(\mathbf{Q}, \mathbf{P}, s) b(\mathbf{Q}, \mathbf{P}, t) \rangle. \quad (4.20)$$

Here $\langle a(s)b(t) \rangle$ is the equilibrium time correlation function between $a(s)$ and $b(t)$. Equation (4.20) is a fundamental result in statistical physics. It reveals that the response of a system to a weak external perturbation driving the system away from equilibrium can be read off in terms of a suitable equilibrium time correlation function. In other words, the properties of a system pulled slightly *out of equilibrium* can be inferred from the knowledge of the statistical properties of the system *at equilibrium*. A seemingly similar relation holds also for the response of a system driven away by a weak perturbation from a nonequilibrium steady state [5].

4.2 Linearized Hydrodynamics and Collective Modes

In this section we aim at deriving an expression for $S_{n,n}(\mathbf{k}, \omega)$ in the hydrodynamic regime. Our starting point will be the NSF equations of hydrodynamics, characterized by the constitutive equations (3.15) and (3.16) for the stress tensor and the heat flux. From the expression for the hydrodynamic fields introduced in Sect. 3.1, we can define the fluctuations of these fields around equilibrium as follows:

$$\rho(\mathbf{r}, t) = \rho_0 + \delta\rho(\mathbf{r}, t),$$
$$\mathbf{u}(\mathbf{r}, t) = \delta\mathbf{u}(\mathbf{r}, t),$$
$$T(\mathbf{r}, t) = T_0 + \delta T(\mathbf{r}, t), \quad (4.21)$$

where we choose a reference frame in which the equilibrium bulk velocity $\mathbf{u}_0$ is zero. The response of the system to the small perturbations resulting from the spontaneous equilibrium fluctuations always present in the system can be described in terms of the linearized hydrodynamic equations. That is, if we retain only the first power of the fluctuations (4.21), the macroscopic equations (3.9), equipped with the NSF constitutive equations (3.15) and (3.16), attain the structure [1]

$$\partial_t \delta\rho = -\rho_0 \nabla_\mathbf{r} \cdot \mathbf{u},$$

$$\rho_0 \partial_t \mathbf{u} = \eta \nabla_\mathbf{r}^2 \mathbf{u} + \left(\frac{1}{3}\eta + \zeta\right) \nabla_\mathbf{r} (\nabla_\mathbf{r} \cdot \mathbf{u}) - c_0^2 \gamma^{-1} \left[\nabla_\mathbf{r}\delta\rho + \rho_0 \alpha \nabla_\mathbf{r}\delta T\right],$$

$$\partial_t \delta T = \frac{\lambda}{\rho_0 c_v} \nabla_\mathbf{r}^2 \delta T - \frac{\gamma - 1}{\alpha} \nabla_\mathbf{r} \cdot \mathbf{u},$$
(4.22)

where $\gamma = c_p/c_v$ is the ratio of the specific heats at constant pressure and constant volume respectively, $\alpha = \rho \, (\partial V/\partial T)_p$ is the coefficient of thermal expansion, and $c_0 = \sqrt{(\partial p/\partial \rho)_s}$ is the adiabatic speed of sound, with the subscript s denoting the entropy per unit mass. When passing into Fourier space, we denote by $\left[\rho_k, \mathbf{u}_k, T_k\right]$ the Fourier transforms of the fluctuations (4.21). It then proves convenient to split the velocity into longitudinal and transversal components lying respectively in the directions parallel and orthogonal to $\mathbf{k}$:

$$\mathbf{u_k}(t) = u^\parallel(\mathbf{k}, t)\hat{\mathbf{k}} + \mathbf{u}^\perp(\mathbf{k}, t),$$

with $\hat{\mathbf{k}}$ denoting a unit wave vector in the direction of $\mathbf{k}$. A relevant property of the linearized NSF equations (4.22) concerns the fact that the time evolution of the transversal components is *decoupled* from that pertaining to the longitudinal components, as we will also see in Chap. 6. To see this, let us introduce first the Laplace transform of $n_\mathbf{k}(t)$, defined as

$$\rho_\mathbf{k}(z) = \int\limits_0^{+\infty} e^{-zt} \rho_\mathbf{k}(t) \mathrm{d}t.$$

By Laplace transforming equation (4.22), one obtains a set of algebraic equations, which can be solved to second order in k. This procedure yields the roots [2]

$$z_1 = -D_T k^2,$$
(4.23)

$$z_{2,3} = \pm i c_0 k - \Gamma k^2,$$
(4.24)

$$z_{4,5} = -\eta_t k^2,$$
(4.25)

where

$$D_T = \lambda/(\rho_0 c_p)$$

is the thermal diffusivity,

$$\Gamma = \frac{1}{2}\left[\eta_\ell + (\gamma - 1)D_T\right]$$

is the sound attenuation coefficient, $\eta_t = \eta/\rho_0$ is the transverse kinetic viscosity, and $\eta_\ell = (\zeta + 4/3\eta)/\rho_0$ is the longitudinal kinetic viscosity. The roots (4.23)–(4.25) are called *hydrodynamic modes*: they describe the evolution of collective fluctuations in a fluid. The two degenerate roots (4.25) correspond to the shear modes, which are decoupled from the other modes. The evolution equation of the two transverse components of the velocity reads

$$\mathbf{u}_{\mathbf{k}}^{\perp}(t) = \mathbf{u}_{\mathbf{k}}^{\perp}(0)e^{-\eta_t k^2 t}. \tag{4.26}$$

Equation (4.26) reveals that any transverse velocity fluctuations in the fluid decay in time and cannot propagate in the fluid. Moreover, according to Eq. (4.26), the short-wavelength disturbances decay faster than the long-wavelength ones. In contrast, the roots (4.23) and (4.24) are longitudinal modes associated to fluctuations of the density, temperature, and longitudinal component of the velocity. The evolution of the fluctuations of the mass density is easily derived:

$$\rho_{\mathbf{k}}(t) = \left[\left(\frac{\gamma - 1}{\gamma} \right) e^{-D_T k^2 t} + \frac{1}{\gamma} e^{-\Gamma k^2 t} \cos(c_0 k t) \right] \rho_{\mathbf{k}}(0). \tag{4.27}$$

From the structure of Eq. (4.27), one can interpret the mode (4.23) as entropy fluctuations at constant pressure that do not propagate in the fluid and give rise to a purely diffusive effect, whose characteristic decay time is $\left(D_T k^2 \right)^{-1}$. By contrast, the two modes (4.24) correspond to sound waves, i.e., pressure fluctuations at constant entropy propagating in the fluid at the speed of sound c_0. Also, these fluctuations fade off eventually, due to the combined effect of viscosity and thermal conduction, with a characteristic time $\left(\Gamma k^2 \right)^{-1}$. From the expression (4.27) for the mass density $\rho_{\mathbf{k}} = m n_{\mathbf{k}}$, one may formally use the procedure outlined in Sect. 4.1 to calculate the dynamic structure factor in the hydrodynamic regime (corresponding to the NSF equations).

A remark is in order here, however.

In the hydrodynamic regime, the phase function $n(\mathbf{Q}, \mathbf{r})$; in (4.1), must be replaced by its average over a mesoscopic cell of linear size ℓ_{meso} [9]. That is, the fluctuations of interest here are macroscopic fluctuations of the particle number density around its thermodynamic equilibrium value. In particular, the length ℓ_{meso} is assumed to be macroscopically small but still sufficiently large to ensure that inside each cell, local equilibrium holds and the relative fluctuation of the number of particles is negligible. It is also worth pointing out that in the hydrodynamic regime, the average (4.12) is no longer an ensemble average, as in (4.1), but corresponds to an average over initial conditions, weighted by the probability density of thermodynamic fluctuation theory [2], which is given by

$$p \propto e^{\Delta S/k_B},$$

where ΔS is the entropy change of the system due to the macroscopic fluctuation. It can thus be shown [2] that in the hydrodynamic regime, $S_{n,n}(\mathbf{k}, \omega)$ attains the form

$$S_{n,n}(\mathbf{k}, \omega) = \frac{1}{2\pi} S_{n,n}(\mathbf{k}) \left[\left(1 - \frac{1}{\gamma}\right) \frac{2D_T k^2}{\omega^2 + \left(D_T k^2\right)^2} + \right.$$
$$\left. + \frac{1}{\gamma} \left(\frac{\Gamma k^2}{(\omega - c_0 k)^2 + \left(\Gamma k^2\right)^2} + \frac{\Gamma k^2}{(\omega + c_0 k)^2 + \left(\Gamma k^2\right)^2} \right) \right]. \quad (4.28)$$

Thus, $S_{n,n}(\mathbf{k}, \omega)$ consists of three components: the Rayleigh line centered at $\omega = 0$, and the two Brillouin lines, located at $\pm c_0 k$; cf. the left panel of Fig. 4.1. As mentioned in Sect. 4.1, the power spectrum constitutes an important source of information for experimentalists. The measure of the Brillouin–Rayleigh spectrum by light scattering in a simple liquid makes it possible, in fact, to compute the speed of sound c_0 (from the position of the Brillouin peaks) as well as the lifetime of the thermal fluctuations and that of the sound waves (from the knowledge of D_T and Γ respectively). A further characterization of the hydrodynamic regime can be obtained by looking at the self part of the van Hove function. It is possible to show [1, 9] that in the limit of small wave vectors and low frequencies, $G_s(\mathbf{r}, t)$ behaves as if the tagged particle were undergoing a diffusive process. The evolution equation is hence of the form

$$\partial_t G_s(\mathbf{r}, t) = -D \nabla_{\mathbf{r}}^2 G_s(\mathbf{r}, t), \quad (4.29)$$

where D is the coefficient of self diffusion. Equation (4.29) admits the solution

$$G_s(\mathbf{r}, t) = \frac{1}{\sqrt{(4\pi Dt)^3}} e^{-\left(\frac{r^2}{4Dt}\right)}. \quad (4.30)$$

By defining the self part $S_{n,n}^{(s)}(\mathbf{k}, \omega)$ of the dynamic structure factor as the double Fourier transform of $G_s(\mathbf{r}, t)$, we obtain from (4.30) the following expression [9]:

$$S_{n,n}^{(s)}(\mathbf{k}, \omega) = \frac{1}{\pi} \frac{Dk^2}{\omega^2 + \left(Dk^2\right)^2}. \quad (4.31)$$

The self-diffusion coefficient D can hence be obtained from $S_{n,n}^{(s)}$ using the relation [10]

$$D = \lim_{\omega \to 0} \lim_{k \to 0} \frac{\pi \omega^2}{k^2} S_{n,n}^{(s)}(\mathbf{k}, \omega), \quad (4.32)$$

where it is crucial that the limits be taken in the indicated order. Equations (4.17) and (4.30) reveal that $G_s(\mathbf{r}, t)$ is a Gaussian function in both the free-particle and hydrodynamic regimes. Consequently, the mean square displacement of the tagged particle, defined as

$$\langle r^2(t) \rangle \equiv \langle |\mathbf{r}(t) - \mathbf{r}(0)| \rangle = \int r^2 G_s(\mathbf{r}, t) d\mathbf{r}, \quad (4.33)$$

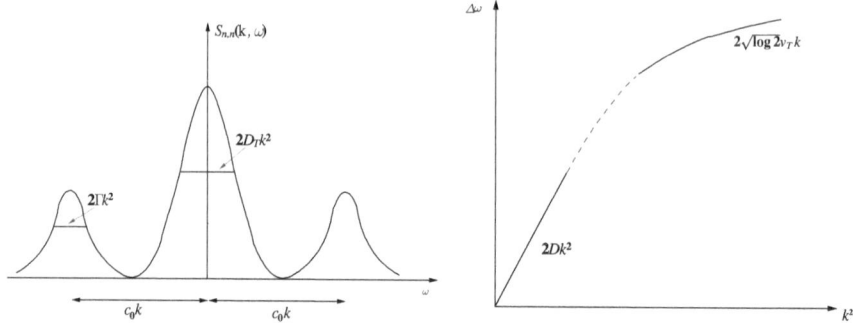

Fig. 4.1 *Left panel.* Dynamical structure factor $S_{n,n}(\mathbf{k}, \omega)$ in the hydrodynamic regime corresponding to the linearized NSF equations. The *central peak* represented in the figure is the Rayleigh peak, which corresponds to thermal fluctuations, whereas the two side peaks are the Brillouin peaks, which originate from pressure fluctuations in the fluid. *Right panel.* Schematic behavior of the full width at half maximum, $\Delta\omega$, of $S_{n,n}^{(n)}(\mathbf{k}, \omega)$. The *straight lines* correspond, respectively, to the hydrodynamic regime (*left branch*) and to the free-particle regime (*right branch*). The *red dashed line* merely represents a possible smooth interpolation between the two limiting regimes [1]

takes, in the free-particle limit, the form

$$\langle r^2(t)\rangle = \frac{3}{2}v_T^2 t^2,$$

which corresponds to a ballistic-like transport, whereas in the hydrodynamic regime, it reads

$$\langle r^2(t)\rangle = \frac{3}{2}Dt,$$

which is the hallmark of a standard diffusion process.

Furthermore, Eq. (4.31) reveals that in the hydrodynamic regime, the full width at half maximum of $S_{n,n}^{(s)}$, denoted by $\Delta\omega$, reads

$$\Delta\omega = 2Dk^2, \quad \text{for } k \ll 1,$$

$$\Delta\omega = 2\sqrt{\ln 2}v_T k, \quad \text{for } k \gg 1.$$

Therefore, $\Delta\omega$ is proportional to k^2 for small wave vectors and is linear in k for large wave vectors, as shown in the right panel of Fig. 4.1. Due to the asymptotic expressions of the self part of the van Hove function for $t \rightarrow 0$ (free-particle limit) and for $t \rightarrow \infty$ (hydrodynamic limit), it would be tempting to stipulate a Gaussian-like behavior of G_s for all times. Nevertheless, molecular dynamics simulations for argon-like liquids [9] reveal that in the intermediate regime, there are deviations of about 10% from the assumed Gaussian behavior. Thus, the challenge of a generalized hydrodynamic theory consists in providing a theoretical framework to interpolate between the two asymptotic regimes (represented by the red dashed line in the right panel of Fig. 4.1). A traditional strategy is based on the introduction of nonlocal hydrodynamic variables and nonlocal transport coefficients, which can be obtained, in a somewhat empirical manner, by solving a generalized Langevin equation [1]. In the next chapter, we will instead learn how the invariant manifold method allows us to extend the hydrodynamic description to the intermediate regime represented in Fig. 4.1. To this end, we will investigate some models of the Boltzmann equation and find an exact closure for the single-particle distribution function.

References

1. J.P. Boon and S. Yip, *Molecular Hydrodynamics* (Dover, 1991).
2. L. E. Reichl, *A modern course in statistical physics* (University of Texas Press, Austin, 1980).
3. U. Marini Bettolo Marconi, A. Puglisi, L. Rondoni and A. Vulpiani, Fluctuation-Dissipation: Response Theory in Statistical Physics, *Phys. Rep.* **461**, 111 (2008).
4. V. Lucarini and M. Colangeli, beyond the linear fluctuation-dissipation theorem: the role of causality, *J. Stat. Mech.* P05013 (2012).
5. M. Colangeli, L. Rondoni and A. Vulpiani, Fluctuation-dissipation relation for chaotic non-hamiltonian systems, *J. Stat. Mech.* L04002 (2012).
6. L. Bertini, A.D. Sole, D. Gabrielli, G. Jona-Lasinio and C. Landim, Macroscopic fluctuation theory for stationary non-equilibrium states, *J. Stat. Phys.* **107**, 635 (2002).
7. L. Onsager and S. Machlup, fluctuations and irreversible processes, *Phys. Rev.* **91**, 1505 (1953).
8. R. Balescu, *Equilibrium and nonequilibrium statistical mechanics* (Wiley, 1975).
9. J.-P. Hansen and I.R. McDonald, *Theory of Simple Liquids* (Academic Press, 2006).
10. D. Forster, *Hydrodynamic fluctuations, Broken Symmetry, and Correlation Functions* (W. A. Benjamin, New York, 1975).
11. R. Kubo, The fluctuation-dissipation theorem, *Rep. Prog. Phys.* **29**, 255 (1966).
12. R. Kubo, Statistical-mechanical theory of irreversible processes. I. General theory and simple applications to magnetic and conduction problems, *J. Phys. Soc. Japan* **12**, 570 (1957).
13. M. Baiesi, C. Maes and B. Wynants, Nonequilibrium linear response for Markov dynamics: I. Jump Processes and Overdamped Diffusion, *J. Stat. Phys.* **137**, 1094 (2009).
14. M. Baiesi, E. Boksenbojm, C. Maes and B. Wynants, Nonequilibrium linear response for Markov dynamics: II. Inertial dynamics, *J. Stat. Phys.* **139**, 492 (2010).
15. M. Colangeli, C. Maes and B. Wynants, A meaningful expansion around detailed balance, *J. Phys. A: Math. Theor.* **44**, 095001 (2011).
16. D. Ruelle, General linear response formula in statistical mechanics, and the fluctuation dissipation theorem far from equilibrium, *Phys. Lett. A* **245**, 220 (1998).

Chapter 5
Hydrodynamic Fluctuations from the Boltzmann Equation

Several solution techniques have been introduced in the literature to obtain approximate solutions of the Boltzmann equation. In particular, the CE method extends the hydrodynamics beyond the NSF approximation in such a way that the decay rate of the next-order approximations (Burnett and super-Burnett) are polynomials of higher order in k [1–3]. In such an extension, relaxation rates may become completely unphysical (amplification instead of attenuation), as first shown by Bobylev [4] for a particular case of Maxwell molecules. Therefore, several regularization techniques have been proposed to restore the thermodynamic admissibility of the generalized hydrodynamic equations [5–7]. A promising route, in particular, is based on the notion of invariant manifold [1], introduced in Sect. 3.4. The method requires a neat separation between hydrodynamic and kinetic (time and length) scales, and postulates the existence of a stable invariant manifold in the space of distribution functions, parameterized by the values of the hydrodynamic fields. Following the approach traced in [8], we will employ here the invariant manifold technique to determine the hydrodynamic modes and the transport coefficients beyond the standard hydrodynamic regime. In particular, we expect to recover the asymptotic form of the dynamic structure factor in the free-particle regime and to shed light on the properties of the hydrodynamic equations in the intermediate regime of finite Knudsen numbers.

This chapter is structured as follows.

We will review, in Sect. 5.1, the eigenvalue problem associated with the linearized Boltzmann equation. In Sect. 5.2, we will derive the invariance equation for the Boltzmann equation equipped with an arbitrary linearized collision operator. In Sect. 5.3, we will introduce a suitable coordinate system that allows us to highlight the symmetries of the solutions of the invariance equation. In Sect. 5.4, we will then solve the invariance equation for the linearized BGK model [9]. In Sect. 5.5, we will investigate the properties of the solution of the invariance equation for a gas of Maxwell molecules. We will therefore clarify the structure of the obtained hydrodynamic modes and cast the generalized transport coefficients in the Green–Kubo

M. Colangeli, *From Kinetic Models to Hydrodynamics*, SpringerBriefs in Mathematics,   49
DOI: 10.1007/978-1-4614-6306-1_5, © Matteo Colangeli 2013

formalism [10]. Finally, we will determine the spectrum of the density fluctuations and discuss some relevant features of the resulting short-wavelength hydrodynamics.

5.1 Eigenfrequencies of the Boltzmann Equation

In this section, we return to the eigenvalue problem for the linearized Boltzmann equation, introduced in Sect. 2.4. Namely, for an inhomogeneous gas, the eigenvalues correspond to k-dependent frequencies (i.e., the inverse of characteristic collision rates). The connection between the fluctuations of the macroscopic variables and the underlying characteristic kinetic rates is a central problem in statistical mechanics [11, 12], one that still lacks a conclusive settlement. In order to appreciate the problem, we anticipate that in the NSF approximations, the hydrodynamic modes are quadratic in the wave vector [13], cf. Eqs. (4.23)–(4.25), and are unbounded. On the other hand, Boltzmann's collision term features equilibration with finite characteristic rates. Hence, the "finite collision frequency" is strongly at variance with the arbitrary decay rates in the NSF approximation: intuitively, the hydrodynamic modes at large k cannot relax faster than the collision frequencies. In his seminal work [14] on the eigenfrequencies of the Boltzmann equation, Resibois provided an explicit connection between the generalized frequencies of the linearized Boltzmann equation and the decay rate of the hydrodynamic fluctuations. He tackled the problem by solving, perturbatively, the eigenvalue problem associated to the Boltzmann equation and to the NSF equations of hydrodynamics.

Let us rewrite the Boltzmann equation (2.40):

$$\partial_t f = -\mathbf{v} \cdot \nabla f + Q(f, f). \tag{5.1}$$

We introduce the dimensionless peculiar velocity $\mathbf{c} = (\mathbf{v} - \mathbf{u}_0)/v_T$ and the equilibrium values of hydrodynamic fields: equilibrium particle number n_0, equilibrium mean velocity $\mathbf{u}_0 = \mathbf{0}$, and equilibrium temperature T_0. The global Maxwellian reads $f^{GM} = (n_0/v_T^3) f_0(c)$, where $f_0(c) = \pi^{-3/2} e^{-c^2}$ is a Gaussian in the velocity space ($c = |\mathbf{c}|$). We linearize (5.1) by considering only small disturbances from the global equilibrium. Moreover, we write the nonequilibrium distribution function (cf. also Table 5.1) as

$$f(\mathbf{r}, \mathbf{c}, t) = f^{LM} + \delta f, \tag{5.2}$$

where f^{LM} denotes the local Maxwellian, to be made precise in Sect. 5.2, and δf is the deviation from local equilibrium. An alternative notation is introduced via $\delta f = f^{GM} \delta \varphi$. We also consider a reference frame moving with the flow velocity and linearize the collision operator around global equilibrium, as shown in Sect. 2.4. When passing to Fourier space, we seek solutions of the form

$$f(\mathbf{r}, \mathbf{c}, t) = e^{\omega t} e^{i \mathbf{k} \cdot \mathbf{r}} f(\mathbf{k}, \mathbf{c}, \omega),$$

Table 5.1 Notation used in this book

f	$=$		f^{LM}		$+$	δf
	$=$	$f^{\text{GM}} \quad +$	$f^{\text{GM}} \varphi_0$		$+$	$f^{\text{GM}} \delta\varphi$
	$=$	$f^{\text{GM}} \quad +$	$f^{\text{GM}} \mathbf{X}^{(0)} \cdot \mathbf{x}$		$+$	$f^{\text{GM}} \delta\mathbf{X} \cdot \mathbf{x}$
	$=$	$f^{\text{GM}} \quad +$	$f^{\text{GM}} \Delta\mathbf{X} \cdot \mathbf{x}$			
	$=$	$f^{\text{GM}} \quad +$	Δf			

Terms have been grouped and abbreviated as depicted here f^{GM} and f^{LM} denote the global and local Maxwellian, respectively, and Δf and δf their "distance" from f. The third row gives information about the closure discussed in this book, while $\mathbf{x}$ is a set of distinguished (lower-order) moments of f

where ω is a complex-valued frequency and k is a real-valued wave vector. Thus, the Boltzmann equation (5.1) reduces to

$$\frac{1}{v_T} \partial_t f(\mathbf{k}, \mathbf{c}, \omega) = -i\mathbf{k} \cdot \mathbf{c} f + \hat{L}\delta f(\mathbf{k}, \mathbf{c}, \omega), \qquad \hat{L} = \frac{1}{v_T} L, \qquad (5.3)$$

where we made use of the fact that $Lf^{\text{LM}} = 0$. In the remainder of this section, we will investigate the spectrum of the operator $\Lambda \equiv \hat{L} - i\mathbf{k} \cdot \mathbf{c}$, which determines the time evolution of the single-particle distribution function [15]. This is readily seen by inspection of the inverse Laplace transform:

$$f(\mathbf{k}, \mathbf{c}, t) = \left[\frac{1}{2\pi i} \oint \frac{e^{zt}}{(z - \Lambda)} dz \right] f(\mathbf{k}, \mathbf{c}, 0), \qquad (5.4)$$

where the closed path encircles the poles of the function inside the integral. According to the spectral theorem, these poles correspond to the spectrum of Λ. The flow term $-i\mathbf{k} \cdot \mathbf{c} f$ is treated here as a small perturbation [16] (this amounts to considering small gradients in the real space). Equation (5.3) can be therefore written in the form

$$\Lambda f = \omega f, \qquad (5.5)$$

which constitutes the starting point of our analysis. We can use here the mathematical framework earlier developed in Sect. 2.4. Thus, we introduce a Hilbert space $\mathscr{H}$ endowed with the scalar product $\langle g|h \rangle$ defined by Eq. (2.65). The analysis of the spectrum of the full operator Λ in Eq. (5.5) requires a preliminary discussion about the properties of the spectrum in the limit $k \to 0$. In this limit, Eq. (5.5) reads

$$\hat{L}\Psi_i(\mathbf{c}) = \lambda_i \Psi_i(\mathbf{c}). \qquad (5.6)$$

The linear operator $\hat{L}$ introduced in Sect. 2.4 is self-adjoint with respect to the scalar product (2.65), whence the corresponding eigenfunctions are (or can be made) orthogonal and constitute a complete set. In particular, a subset of them spanning a five-dimensional subspace of $\mathscr{H}$ is related to the fivefold degenerate zero eigenvalue. These functions correspond to the collision invariants $f^{GM}\mathbf{X}^{(0)}$, with $\mathbf{X}^{(0)}$ denoting a set of lower-order Sonine (or associated Laguerre) polynomials:

$$\mathbf{X}^{(0)} = \left[1, 2\mathbf{c}, \left(c^2 - \frac{3}{2}\right)\right]. \tag{5.7}$$

One also typically assumes that the eigenvalues of $\hat{L}$ other than 0 have no accumulation point at the origin. This assumption is always implicit in any calculation of transport coefficients based on kinetic theory. Physically, it implies a separation of the relaxation time scale λ_i^{-1} and the hydrodynamic time scale corresponding to $(v_t k)^{-1}$ occurring in Eq. 4.18. The lack of such a time scale separation can cause the breaking of the hydrodynamic description, as will be discussed in Chap. 6, for a Grad's moment system. For finite values of k, there will be a set of eigenvalues of Λ, denoted by ω_α, with $\alpha = 1, \ldots, 5$, which in the $k \to 0$ limit, reduce to the aforementioned degenerate zero eigenvalue. Hence, we briefly review here the results of a perturbative method, outlined in [14], that allows us to determine the dependence of the set ω_α on k. If we denote by $\Psi_\alpha(k)$ the eigenfunctions corresponding to $\omega_\alpha(k)$, the yet unknown eigenfunctions and eigenvalues can be expanded in powers of the wave vector k:

$$\Psi_\alpha = \Psi_\alpha^{(0)} + k\Psi_\alpha^{(1)} + k^2\Psi_\alpha^{(2)} + \cdots,$$
$$\omega_\alpha = \omega_\alpha^{(0)} + k\omega_\alpha^{(1)} + k^2\omega_\alpha^{(2)} + \cdots, \tag{5.8}$$

where $\Psi_\alpha^{(0)}$ are linear combinations of the collision invariants (5.7), whose detailed expression is not relevant here (cf. [17] for details). The use of a Rayleigh–Schrödinger perturbation theory leads to the following polynomial expression for the set $\{\omega_\alpha\}$ [17]:

$$\omega_1 = ic_0 k - k^2 \left\langle \Psi_1^{(0)} \left| \left((c_x - c_0)\frac{1}{\hat{L}}(c_x - c_0)\right) \Psi_1^{(0)} \right\rangle,$$
$$\omega_2 = -ic_0 k - k^2 \left\langle \Psi_2^{(0)} \left| \left((c_x + c_0)\frac{1}{\hat{L}}(c_x + c_0)\right) \Psi_2^{(0)} \right\rangle,$$
$$\omega_3 = -k^2 \left\langle \Psi_3^{(0)} \left| \left(c_x \frac{1}{\hat{L}} c_x\right) \Psi_3^{(0)} \right\rangle,$$
$$\omega_4 = -k^2 \left\langle \Psi_4^{(0)} \left| \left(c_x \frac{1}{\hat{L}} c_x\right) \Psi_4^{(0)} \right\rangle,$$
$$\omega_5 = -k^2 \left\langle \Psi_5^{(0)} \left| \left(c_x \frac{1}{\hat{L}} c_x\right) \Psi_5^{(0)} \right\rangle, \tag{5.9}$$

where c_0 is the speed of sound, defined in Sect. 4.2.

On the other hand, in Fourier space, the linearized NSF equations (4.22) read

$$\omega n_k = -i n_0 \left(\mathbf{k} \cdot \mathbf{u}_k \right),$$

$$\omega \mathbf{u}_k = -\eta k^2 \mathbf{u}_k - \left(\frac{1}{3}\eta + \zeta \right) \left(\mathbf{k} \cdot \mathbf{u}_k \right) \mathbf{k} - i c_0^2 \gamma^{-1} \rho_0^{-1} \mathbf{k} n_k - i c_0^2 \gamma^{-1} \alpha \mathbf{k} T_k,$$

$$\omega T_k = -\frac{\lambda}{\rho_0 c_v} k^2 T_k - i \frac{\gamma - 1}{\alpha} \left(\mathbf{k} \cdot \mathbf{u}_k \right). \tag{5.10}$$

The condition of nontrivial solvability of the linear system (5.10) with respect to the variables $[n_k(\omega), \mathbf{u}_k(\omega), T_k(\omega)]$ yields the *dispersion relation* $\omega(k)$, i.e., the normal mode frequencies of the system. The roots of the dispersion relation enjoy the same structure as that of the solutions (4.23)–(4.25). In particular, the real part of these modes is quadratic in the wave vector [1, 14, 17]:

$$\mathrm{Re}(\omega) \propto -k^2,$$

which is a hallmark of the NSF approximation. Postulating the equivalence of the hereby obtained hydrodynamic frequencies with the set of kinetic frequencies ω_α given by Eq. (5.9) leads to approximate expressions for the transport coefficients, which can be shown to be equivalent to the reduced expressions determined by many-body autocorrelation functions [14]. As also pointed out in [16], the result obtained by Resibois based on the correspondence between hydrodynamic modes and kinetic frequencies reveals that in the limit of long wavelengths, the possible modes of motion of the gas correspond to rather ordered motions, such as propagation of a sound wave. These modes are referred to in the literature as *collective* modes, because they involve the coordinate action of a huge number of particles. The onset of such an ordered motion as a result of the underlying chaoticness of the individual motion of the particles is a striking feature of statistical mechanics. The reason for this can be traced back to the effect of the collisions, which very quickly drive the system toward the local equilibrium state, which is a highly organized one. From then onward, the flow term produces slow variations in space and time of this basic state, which reduce the local gradients of the hydrodynamic fields; cf. also the discussion at the end of Chap. 3. In the sequel of this chapter, we will employ invariant manifold theory to obtain a generalization of the pioneering approach developed by Resibois.

5.2 The Invariant Manifold Technique

Let us briefly recall some basic mathematical tools of the method introduced in Chap. 3. We denote by U and $\mathbf{x}(\mathbf{r}, t)$, respectively, the space of single-particle distribution functions $f(\mathbf{r}, \mathbf{v}, t)$ and some of its distinguished moments. We define the *locally finite-dimensional* manifold $\Omega \subset U$ as the set of functions $f(\mathbf{x}(\mathbf{r}, t), \mathbf{c})$ whose dependence on the space–time variables $(\mathbf{r}, t)$ is parameterized through $\mathbf{x}(\mathbf{r}, t)$. In this chapter, we will identify the distinguished moments $\mathbf{x}(\mathbf{r}, t)$ with the hydrodynamic fields. Moreover, P is the thermodynamic projection operator, which, as discussed in Sect. 3.4, allows us to decompose the dynamics into a fast motion on the affine subspace $f + \ker[P]$ and a slow motion, which occurs along the tangent space T_f. The use of the thermodynamic projector guarantees the persistence of dissipation: it can be shown [1] that the entropy production rate is unaltered when the dynamic is projected along the manifold of slow motion.

In order to derive exact hydrodynamic equations from the general eigenvalue problem (5.3), we proceed as follows:

1. We determine the invariant manifold, i.e., the distribution function solving the invariance equation

$$(\mathbf{1} - P)\Lambda\Delta f = 0, \tag{5.11}$$

   where $\Delta f \equiv f - f^{GM}$ (cf. also Table 5.1).
2. We derive the equations of linear hydrodynamics by integrating the kinetic equation (5.3), with f given by the solution of Eq. (5.11). By construction, the hydrodynamic modes then coincide with the set ω_α of eigenfrequencies of the Boltzmann equation, which vanish in the limit $k \to 0$.

We also denote by $\mathbf{x}_k$ the Fourier components of the dimensionless hydrodynamic fluctuations $[\tilde{n}, \tilde{\mathbf{u}}, \tilde{T}]$: $\tilde{n} \equiv (n - n_0)/n_0$ (particle number perturbation), $\tilde{\mathbf{u}} \equiv \mathbf{u}/v_T$ (velocity perturbation), and $\tilde{T} \equiv (T - T_0)/T_0$ (temperature perturbation). Further, we split the mean velocity $\tilde{\mathbf{u}}$ uniquely as $\tilde{\mathbf{u}} = u^\| \mathbf{e}_\| + u^\perp \mathbf{e}_\perp$, where the unit vector $\mathbf{e}$ is parallel to $\mathbf{k}$, and $\mathbf{e}_\perp$ is orthonormal to $\mathbf{e}_\|$, i.e., $\mathbf{e}_\perp$ lies in the plane perpendicular to $\mathbf{k}$. Due to isotropy, $u^\perp$ alone fully represents the twice degenerated (shear) dynamics. By linearizing around the global equilibrium, we write the local Maxwellian contribution to f in (5.2) as $f^{LM} = f^{GM}(1 + \varphi_0)$, where φ_0 takes a simple form, $\varphi_0 = \mathbf{X}^{(0)} \cdot \mathbf{x}$ (linear quasiequilibrium manifold), where $\mathbf{X}^{(0)}(\mathbf{c})$ was defined in Eq. (5.7). It is conveniently considered a four-dimensional vector using the four-dimensional version $\mathbf{x}_k = [\tilde{n}_k, u^\|, \tilde{T}_k, u^\perp]$, and is then given by (5.21). It proves convenient to introduce a vector of velocity polynomials $\boldsymbol{\xi}(\mathbf{c})$, which is similar to $\mathbf{X}^0$ and defined below in Eq. (5.22), such that using the notation introduced in Eq. (2.65),

$$\langle \xi_\mu | X_\nu^{(0)} \rangle = \delta_{\mu\nu}.$$

Hence the fields $\mathbf{x}_k$ are obtained as $\langle \boldsymbol{\xi}(\mathbf{c}) \rangle_{f^{LM}} = \mathbf{x}_k$, where averages are defined here as

$$\langle \boldsymbol{\xi}(\mathbf{c}) \rangle_f = \frac{1}{n_0} \int \boldsymbol{\xi}(\mathbf{c}) f(\mathbf{c}) d\mathbf{v}. \tag{5.12}$$

We introduce yet unknown fields $\delta\mathbf{X}(\mathbf{c}, \mathbf{k})$ that characterize the part δf of the distribution function. As long as deviations from the local Maxwellian are small, we seek a nonequilibrium manifold that is also linear in the hydrodynamic fields $\mathbf{x}$ themselves. Therefore, we set

$$\delta\varphi = \delta\mathbf{X} \cdot \mathbf{x}_k. \tag{5.13}$$

The "eigen"closure (5.13), which formally and very generally addresses the fact that we wish to *not* include other than hydrodynamic variables, implies a closure between moments of the distribution function, to be worked out in detail below. If we use the above form (5.13) for $\delta f = f^{\text{GM}} \delta \varphi$, with $\hat{L} \delta f = f^{\text{GM}} L[\delta \mathbf{X}] \cdot \mathbf{x}_k$, and the canonical abbreviations $\Delta \mathbf{X} \equiv \mathbf{X}^0(\mathbf{c}) + \delta \mathbf{X}(\mathbf{c}, \mathbf{k})$, then Eq. (5.5) reads

$$\omega f^{\text{GM}} \Delta \mathbf{X} \cdot \mathbf{x}_k = \Lambda \Delta f = -i\mathbf{k} \cdot \mathbf{c} f^{\text{GM}} \Delta \mathbf{X} \cdot \mathbf{x}_k + f^{\text{GM}} \hat{L} \delta \mathbf{X} \cdot \mathbf{x}_k. \qquad (5.14)$$

The microscopic projected dynamics is obtained through the projector P, which, when acting on the vector field $J(f) = \Lambda \Delta f$, gives

$$P \Lambda \Delta f = D_{\mathbf{x}_k} \Delta f \cdot \int \boldsymbol{\xi}(\mathbf{c}) \Lambda \Delta f \mathrm{d}\mathbf{v}, \qquad (5.15)$$

where $D_{\mathbf{x}_k} \Delta f \equiv \partial \Delta f / \partial \mathbf{x}_k$ and the quantity inside the integral in (5.15) represents the time evolution equations for the moments $\mathbf{x}_k$. These are readily obtained by integration of the weighted (5.5) as

$$\omega \langle \boldsymbol{\xi}(\mathbf{c}) \rangle_f = -i\mathbf{k} \cdot \langle \boldsymbol{\xi}(\mathbf{c}) \mathbf{c} \rangle_f + \langle \boldsymbol{\xi}(\mathbf{c}) \rangle_{\hat{L} \delta f}. \qquad (5.16)$$

Due to the eigenclosure (5.13), cf. also Table 5.1, one obtains $D_{\mathbf{x}_k} \Delta f = f^{\text{GM}} \Delta \mathbf{X}$, whereas (5.16) is linear in $\mathbf{x}_k$ and can be written as

$$\omega \mathbf{x}_k = \mathbf{M} \cdot \mathbf{x}_k. \qquad (5.17)$$

Equation (5.17) defines the matrix $\mathbf{M}$ of hydrodynamic coefficients, whose explicit structure will be made clear in Eq. (5.27). Using (5.17), Eq. (5.15) can be cast in the form

$$P \Lambda \Delta f = f^{\text{GM}} \Delta \mathbf{X} \cdot \mathbf{M} \cdot \mathbf{x}_k. \qquad (5.18)$$

In the derivation of (5.18), one needs to take into account the constraints $\langle \boldsymbol{\xi}(\mathbf{c}) \rangle_{\delta f} = \mathbf{0}$ (because the fields $\mathbf{x}_k$ are defined through the local Maxwellian part of the distribution function only) and $\langle \boldsymbol{\xi}(\mathbf{c}) \rangle_{\hat{L} \delta f} = \mathbf{0}$ (conservation laws). The dependence of the matrix elements of $\mathbf{M}$ on moments of δf is explicitly given in Table 5.2. Combining (5.14) and (5.18) and requiring that the result hold for every $\mathbf{x}_k$ (invariance condition), we obtain a closed, singular integral equation (the invariance equation) for complex-valued $\delta \mathbf{X}$,

$$\Delta \mathbf{X} \cdot \mathbf{M} = -i\mathbf{k} \cdot \mathbf{c} \, \Delta \mathbf{X} + \hat{L} \delta \mathbf{X}. \qquad (5.19)$$

Notice that $\delta \mathbf{X} = \Delta \mathbf{X} - \mathbf{X}^{(0)}$ vanishes for $k = 0$, which implies that in the limit $k \to 0$, the invariant manifold is given by the set of local Maxwellians f^{LM}. The implicit Eq. (5.19) for $\delta \mathbf{X}$ (or $\Delta \mathbf{X}$, since $\mathbf{X}^{(0)}$ is known) is identical to the eigenclosure (5.13), and is our main and practically useful result.

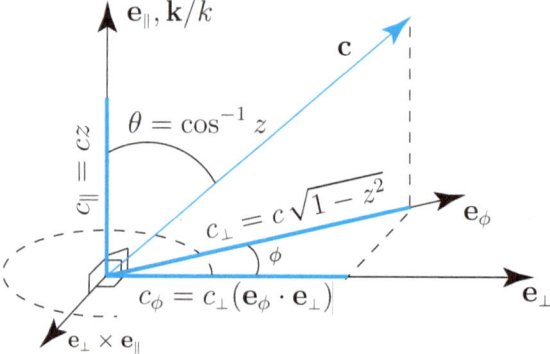

Fig. 5.1 Schematic drawing introducing an orthonormal frame $\mathbf{e}_\|$, $\mathbf{e}_{\|\perp}$, and $\mathbf{e}_{\|\perp} \times \mathbf{e}_\|$, which is defined by the wave vector $\mathbf{k} \parallel \mathbf{e}_\|$ and the heat flux $\mathbf{q}$ (not shown), which lies in the $\mathbf{e}_\| - \mathbf{e}_{\|\perp}$-plane. Shown is the velocity vector $\mathbf{c}$ (5.20) relative to this frame (characterized by length c, coordinate z, and angle ϕ) and its various components. The integration over $d\mathbf{c} = c^2 dc dz d\phi$ is done in spherical coordinates with respect to the local orthonormal basis

5.3 Coordinate Representation and Symmetries

In order to calculate the averages occurring in Sect. 5.2, we switch to spherical coordinates. For each (at present arbitrary) wave vector $\mathbf{k} = k\mathbf{e}_\|$, we choose the coordinate system in such a way that its (vertical) z-direction aligns with $\mathbf{e}_\|$ and its x-direction aligns with $\mathbf{e}_\perp$. The velocity vector was decomposed earlier as $\tilde{\mathbf{u}} = u^\| \mathbf{e}_\| + u^\perp \mathbf{e}_\perp$. We can then express $\mathbf{c}$, over which we are going to perform all integrals, in terms of its norm c, a vertical variable z, and plane vector $\mathbf{e}_\phi$ (azimuthal angle $\mathbf{e}_\phi \cdot \mathbf{e}_\perp = \cos\phi$; the plane contains $\mathbf{e}_\perp$) for the present purpose as

$$\mathbf{c}/c = \sqrt{1 - z^2}\, \mathbf{e}_\phi + z\mathbf{e}_\|, \tag{5.20}$$

as shown in Fig. 5.1.

The local Maxwellian, linearized around the global equilibrium, takes the form $f^{\mathrm{LM}}/f^{\mathrm{GM}} = 1 + \varphi_0 = 1 + \mathbf{X}^{(0)} \cdot \mathbf{x}_k$, where the four-dimensional $\mathbf{X}^{(0)}$ and the related vector $\boldsymbol{\xi}$ employing four-dimensional $\mathbf{x}_k = [\tilde{n}_k, u^\|, \tilde{T}_k, u^\perp]$ are given by the expressions

$$\mathbf{X}^{(0)}(\mathbf{c}) = \left(1, 2c_\|, \left(c^2 - \frac{3}{2}\right), 2c_\perp\right), \tag{5.21}$$

$$\boldsymbol{\xi}(\mathbf{c}) = \left(1, c_\|, \frac{2}{3}\left(c^2 - \frac{3}{2}\right), c_\perp\right), \tag{5.22}$$

which clearly resemble the expression given in (2.71). Here we have introduced, for later use, the abbreviations

$$c_{\parallel} \equiv \mathbf{c} \cdot \mathbf{e}_{\parallel}, \qquad c_{\perp} \equiv \mathbf{c} \cdot \mathbf{e}_{\perp}, \qquad c_{\phi} \equiv \mathbf{c} \cdot \mathbf{e}_{\phi} = \frac{c_{\perp}}{\mathbf{e}_{\perp} \cdot \mathbf{e}_{\phi}}, \qquad (5.23)$$

such that $i\mathbf{k} \cdot \mathbf{c} = ikc_{\parallel}$. We can then rewrite (5.20) as $\mathbf{c} = c_{\phi}\mathbf{e}_{\phi} + c_{\parallel}\mathbf{e}_{\parallel}$ with $c_{\parallel} = cz$ and $c_{\phi} = c\sqrt{1 - z^2}$. The latter two components, contrasted by $c_{\perp}$ (and $\mathbf{e}_{\phi}$), do not depend on the azimuthal angle. We further introduced yet unknown fields $\delta\mathbf{X}(\mathbf{c}, \mathbf{k})$, which characterize the nonequilibrium part of the distribution function $\delta\varphi = \delta f / f^{\mathrm{GM}}$.

By analogy with the structure of the local Maxwellian, we postulate that close to equilibrium, $\delta\varphi$ depends linearly on the hydrodynamic fields $\mathbf{x}_k$ themselves. Equation (5.13) can therefore be cast in the form

$$\delta\varphi = \delta\mathbf{X} \cdot \mathbf{x}_k = \delta X_1 \tilde{n}_k + \delta X_2 u^{\parallel} + \delta X_3 \tilde{T}_k + \delta X_4 u^{\perp}. \qquad (5.24)$$

The functions $\delta X_{1,2,3}$, which are associated with the longitudinal fields, inherit the full rotational symmetry of the corresponding Maxwellian components, i.e., $\delta X_{1,2,3} = \delta X_{1,2,3}(c, z)$, whereas δX_4 factorizes as $\delta X_4(c, z, \phi) = 2\delta Y_4(c, z)$ $\sum_{m=1}^{\infty} y_m \cos m\phi$. In this context, it is an important technical aspect of our derivation to work with a suitable orthogonal set of basis functions to represent δf uniquely.[1] The matrix $\mathbf{M}$ in (5.18) contains the nonhydrodynamic fields: the heat flux $\mathbf{q}_k \equiv \left\langle \mathbf{c}\left(c^2 - \frac{5}{2}\right)\right\rangle_f$ and the stress tensor $\sigma_k \equiv \langle \overline{\mathbf{cc}} \rangle_f$, where $\overline{\mathbf{s}}$ denotes the symmetric traceless part of a tensor $\mathbf{s}$ [18], $\overline{\mathbf{s}} = \frac{1}{2}(\mathbf{s} + \mathbf{s}^T) - \frac{1}{3}\mathrm{tr}(\mathbf{s})\mathbf{I}$, where $\mathbf{I}$ is the identity matrix. Using (5.13) and the above-mentioned angular dependence of the $\delta\mathbf{X}$ functions (the only term in δX_4 playing a role in our calculations is the first-order term

---

[1] In the notation in [8, 18, 19], the distribution function is written as a sum over n-fold contracted products of nth-rank tensors $f(\mathbf{c}) = f_0(c) \sum_{k,n=0}^{\infty} \langle \phi_k^n \rangle \odot^n \phi_k^n(\mathbf{c})$ with $\langle \phi_k^n \rangle = \int f(\mathbf{c})\phi_k^n \, d\mathbf{c}$ and base functions $\phi_k^n(\mathbf{c}) = l_k^n L_k^{n+1/2}(c^2) \overline{\otimes^n \mathbf{c}}$, where L_k^n are the associated Laguerre (kth-order) polynomials [20], $\otimes^n \mathbf{c}$ denotes the n-fold tensor product, and $\overline{\mathbf{a}}$ denotes the irreducible part of a tensor $\mathbf{a}$. For the explicit construction of nth-rank irreducible tensors $\overline{\otimes^n \mathbf{c}}$, see [18, p. 160]. The normalization coefficients evaluate as $l_k^n = (\sqrt{\pi}k!(1 + 2n)!!/[2(k + n + 1/2)!n!])^{1/2}$. The base function $\phi_k^n(\mathbf{c})$ is thus a $(2k + n)$th-order polynomial in c. The lowest-order base functions read $\phi_0^0 = 1$, $\phi_0^1 = \sqrt{2}\mathbf{c}$, $\phi_1^0 = \sqrt{2/3}(3/2 - c^2)$, $\phi_1^1 = (2/\sqrt{5})(5/2 - c^2)\mathbf{c}$, and $\phi_0^2 = \sqrt{2}\overline{\mathbf{cc}}$. Density, velocity, temperature, heat flux, and stress tensor are related to the moments as follows: $\tilde{n} = \langle \phi_0^0 \rangle$, $\tilde{\mathbf{u}} = \langle \phi_0^1 \rangle / \sqrt{2}$, $\tilde{T} = \langle \phi_1^0 \rangle \sqrt{3/2}$, $\mathbf{q} = \langle \phi_1^1 \rangle$, and $\sigma = \langle \phi_0^2 \rangle / \sqrt{2}$. The distribution function is then split into (orthogonal) parts as $f(\mathbf{c}) = f^{\mathrm{LM}}(\mathbf{c}) + \delta f^{\mathrm{Grad}}(\mathbf{c}) + \delta f^{\mathrm{rest}}(\mathbf{c})$ with $f^{\mathrm{LM}}(\mathbf{c}) \equiv f_0(c)(\langle \phi_0^0 \rangle \phi_0^0 + \langle \phi_0^1 \rangle \phi_0^1 + \langle \phi_1^0 \rangle \phi_1^0)$ and $\delta f^{\mathrm{Grad}}(\mathbf{c}) \equiv f_0(c)(\langle \phi_1^1 \rangle \phi_1^1 + \langle \phi_0^2 \rangle \phi_0^2)$, while the sum in $\delta f^{\mathrm{rest}}(\mathbf{c}) = \sum_{k,n} \langle \phi_k^n \rangle \odot^n \phi_k^n(\mathbf{c})$ extends over the remaining (k, n)-pairs. Number density, velocity, and temperature are therefore determined by f^{LM} alone, and δf automatically obeys constraints such as the orthogonality requirement $\int \delta f(\mathbf{c})\phi_1^0 \, d\mathbf{c} = 0$ and also $\int \delta f(\mathbf{c})\xi(\mathbf{c})d\mathbf{c} = 0$, as mentioned in the text. These conditions become redundant if calculations are performed using the particular basis ϕ_k^n. For Maxwell molecules, the dependence on the polar angle ϕ can be included by replacing $P_l(z)$ by $e^{im\phi} P_l^m(z)$, involving the associated Legendre polynomials [20], and the eigenvalues are independent of m. Then these base functions reduce to the eigenfunctions $\psi_{r,l}(c, z)$ (5.29) of the Maxwell gas.

Table 5.2 Symmetry-adapted components of (nonequilibrium) stress tensor σ_k and heat flux $\mathbf{q}_k$, introduced in (5.25) and (5.26)

$\sigma_1^{\parallel}$	$\sigma_2^{\parallel}$	$\sigma_3^{\parallel}$	σ_4
$\langle \lambda^{\parallel} \delta X_1 \rangle$	$\langle \lambda^{\parallel} \delta X_2 \rangle$	$\langle \lambda^{\parallel} \delta X_3 \rangle$	$\langle c_{\parallel} c_{\perp} \delta Y_4 \rangle$
$-k^2 B$	ikA	$-k^2 C$	ikD
real, $\oplus$	imag., $\oplus$	real, $\oplus$	imag., $\ominus$
$q_1^{\parallel}$	$q_2^{\parallel}$	$q_3^{\parallel}$	q_4
$\langle \gamma^{\parallel} \delta X_1 \rangle$	$\langle \gamma^{\parallel} \delta X_2 \rangle$	$\langle \gamma^{\parallel} \delta X_3 \rangle$	$\left\langle \left(c^2 - \tfrac{5}{2}\right) c_{\perp} \delta Y_4 \right\rangle$
ikX	$-k^2 Z$	ikY	$-k^2 U$
imag., $\ominus$	real, $\ominus$	imag., $\ominus$	real, $\oplus$

Row 2 Microscopic expression of these components (averaging with the global Maxwellian). Short-hand notation used: $\lambda^{\parallel} = c_{\parallel}^2 - \frac{c^2}{3}$ and $\gamma^{\parallel} = \left(c^2 - \frac{5}{2}\right) c_{\parallel}$. *Row 3* Expression of the components in terms of (as we show, real-valued) functions A–Z (see text). *Row 4* Parity with respect to z—symmetric ($\oplus$) or antisymmetric ($\ominus$)—of the part of the corresponding δX entering the averaging in row 2, and whether this part is imaginary or real-valued (see Fig. 5.5). Row 3 is an immediate consequence of row 4

$\cos \phi$, with $y_1 = 1$),[2] constraints such as the required decoupling between longitudinal and transversal dynamics of the hydrodynamic fields are automatically dealt with correctly in performing integrals over ϕ.

More explicitly, the stress tensor and heat flux uniquely decompose as follows:

$$\boldsymbol{\sigma}_k = \sigma^{\parallel} \frac{3}{2} \overline{\mathbf{e}_{\parallel} \mathbf{e}_{\parallel}} + \sigma^{\perp} 2 \overline{\mathbf{e}_{\parallel} \mathbf{e}_{\parallel}}_{\perp}, \tag{5.25}$$

$$\mathbf{q}_k = q^{\parallel} \mathbf{e}_{\parallel} + q^{\perp} \mathbf{e}_{\parallel \perp}, \tag{5.26}$$

with the moments $\sigma^{\parallel} = (\sigma_1^{\parallel}, \sigma_2^{\parallel}, \sigma_3^{\parallel}) \cdot (\tilde{n}_k, u^{\parallel}, \tilde{T}_k)$ and $\sigma^{\perp} = \sigma_4 u^{\perp}$, and similarly for $\mathbf{q}_k$ (see row 2 of Table 5.2).

The prefactors arise from the identities $\overline{\mathbf{e}_{\parallel} \mathbf{e}_{\parallel}} : \overline{\mathbf{e}_{\parallel} \mathbf{e}_{\parallel}} = \frac{2}{3}$ and $\overline{\mathbf{e}_{\parallel} \mathbf{e}_{\parallel}}_{\perp} : \overline{\mathbf{e}_{\parallel} \mathbf{e}_{\parallel}}_{\perp} = \frac{1}{2}$. We note in passing that while the stress tensor has, in general, three different eigenvalues, in the present symmetry-adapted coordinate system, it exhibits a vanishing first normal stress difference. Since the integral kernels of all moments in (5.25) do

---

[2] The integrals listed in Table 5.2 obey the following decoupling rules:

$$\int \left(c_{\parallel}^2 - \frac{1}{3}c^2\right) \delta X_n d\mathbf{c} \propto 1 - \delta_{n,4},$$

$$\int c_{\parallel} c_{\perp} \delta X_n d\mathbf{c} \propto \delta_{n,4},$$

$$\int c_{\parallel} \left(c^2 - \frac{5}{2}\right) \delta X_n d\mathbf{c} \propto 1 - \delta_{n,4},$$

$$\int c_{\perp} \left(c^2 - \frac{5}{2}\right) \delta X_n d\mathbf{c} \propto \delta_{n,4}.$$

not depend on the azimuthal angle, these are actually two-dimensional integrals over $c \in [0, \infty]$ and $z \in [-1, 1]$, each weighted by a component of $2\pi c^2 f^{GM} \delta \mathbf{X}$.

Stress tensor and heat flux can also be written in an alternative form, defined by row 3 of Table 5.2, in terms of the functions A–Z, which correspond to moments of the nonequilibrium distribution function and are related to the generalized transport coefficients; see [8, 13, 21] and below.

Due to fundamental symmetry considerations, the generalized transport coefficients A–Z introduced here are real-valued. To show this, we use the functions A–Z to split $\mathbf{M}$ into parts as $\mathbf{M} = \mathrm{Re}(\mathbf{M}) - i\,\mathrm{Im}(\mathbf{M})$,

$$\mathbf{M} = k^2 \begin{pmatrix} 0 & 0 & 0 & 0 \\ 0 & A & 0 & 0 \\ \frac{2}{3}X & 0 & \frac{2}{3}Y & 0 \\ 0 & 0 & 0 & D \end{pmatrix} - ik \begin{pmatrix} 0 & 1 & 0 & 0 \\ \tilde{B} & 0 & \tilde{C} & 0 \\ 0 & \tilde{Z} & 0 & 0 \\ 0 & 0 & 0 & 0 \end{pmatrix}, \tag{5.27}$$

with abbreviations $\tilde{B} \equiv \dfrac{1}{2} - k^2 B$, $\tilde{C} \equiv \dfrac{1}{2} - k^2 C$, and $\tilde{Z} \equiv \dfrac{2}{3}(1 - k^2 Z)$. The checkerboard structure of the matrix $\mathbf{M}$ (5.27) is particularly useful for studying properties of the hydrodynamic equation (5.18), such as hyperbolicity and stability [2, 3], once the functions A–Z are explicitly evaluated. Moreover, we remind the reader that we use orthogonal basis functions (irreducible moments; cf. Table 5.2) to solve (5.19). In order to show how the above functions enter the definition of the $\mathbf{M}$ matrix, we first notice that its elements are a priori complex-valued. We wish, then, to make use of the fact that all integrals over z vanish for odd integrands. To this end, we introduce abbreviations $\oplus$ ($\ominus$) for a real-valued quantity that is even (odd) with respect to the transformation $z \to -z$. One notices that $\mathbf{X}^{(0)} = (\oplus, \ominus, \oplus, \oplus)$, and we recall that A–Z are integrals over either even or odd functions in z, times a component of $\delta \mathbf{X}$ (see Table 5.2). Let us prove the consistency of the specified symmetry of $\mathbf{M}$ and the invariance condition: start by assuming A–Z to be real-valued functions. Then $M_{\mu\nu} = \oplus$ if $\mu + \nu$ is even, and $M_{\mu\nu} = i\oplus$ otherwise. This implies $\delta X_1 = \oplus + i\ominus$, $\delta X_2 = \ominus + i\oplus$, $\delta X_3 = \oplus + i\ominus$, and $\delta X_4 = \oplus + i\ominus$, i.e., different symmetry properties for real and imaginary parts. With these "symmetry" expressions for $\mathbf{X}^{(0)}$, $\delta \mathbf{X}$, and $\mathbf{M}$ at hand, and by noticing that symmetry properties for $\delta \mathbf{X}$ carry over to $\hat{L}(\delta \mathbf{X})$ because the $\psi_{r,l}$ are (i) symmetric (antisymmetric) in z for even (odd) l and (ii) eigenfunctions of $\hat{L}$, we can insert into the right-hand side of the equation $\hat{L}(\delta \mathbf{X}) = (\mathbf{X}^{(0)} + \delta \mathbf{X}) \cdot (\mathbf{M} + i\ominus \mathbf{I})$, which is identical to the invariance equation (5.19). There are only two cases to consider, because $\mathbf{M}$ has a checkerboard structure, i.e., only two types of columns: columns $\mu = 1$ and $\mu = 3$; here we have $\delta X_\mu = \oplus + i\ominus$ because $M_{1-3,4} = 0$; columns $\mu \in \{2, 4\}$; here $\delta X_\mu = \oplus + i\ominus$ if $M_{\mu,1-3} = 0$ (which is the case for column 4) and $\ominus + i\oplus$ if $M_{\mu,4} = 0$ (which is the case for column 2). These observations complete the proof.

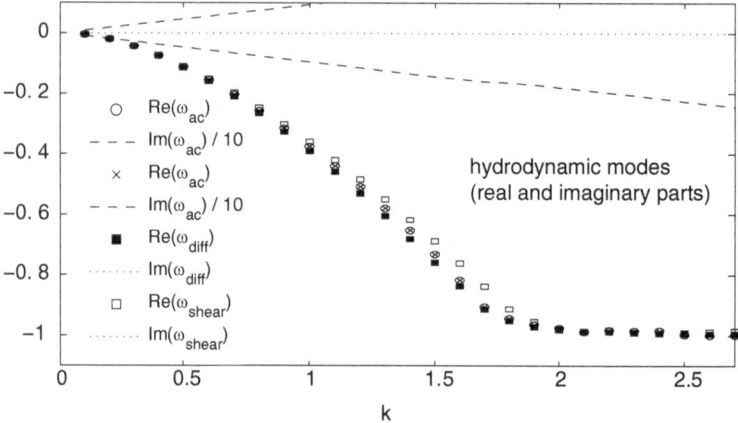

Fig. 5.2 Exact hydrodynamic modes ω of the Boltzmann–BGK kinetic equation as a function of wave number k (two complex-conjugate acoustic modes ω_{ac}, twice degenerated shear mode ω_{shear}, and thermal diffusion mode ω_{diff}). The nonpositive decay rates $\text{Re}(\omega)$ attain the limit of collision frequency (-1) in the $k \gg 1$ regime

5.4 The BGK Kinetic Model

The solution of the invariance equation (5.19) can be obtained in some simple cases amenable to an analytic or numeric treatment. In this section, we focus on the linearized version of the BGK kinetic model (cf. Eq. 2.69), which remains popular in applications [22] and is characterized by a single collision frequency. The invariance equation for this model is readily obtained from Eq. (5.19) by using $\hat{L}(\delta\mathbf{X}) = -\delta\mathbf{X}$, which therefore yields

$$\delta\mathbf{X} = \mathbf{X}^{(0)} \cdot \left(\mathbf{M} + [ikc_{\parallel} + 1]\mathbf{I}\right)^{-1} - \mathbf{X}^{(0)}. \tag{5.28}$$

Notice that $\delta\mathbf{X}$ vanishes for $k = 0$, and that (5.28) is supplemented with the basic constraint $\langle\boldsymbol{\xi}\rangle_{\delta f} = 0$, which, however, is automatically dealt with if we evaluate only anisotropic (irreducible) moments with δf, such as those listed in Table 5.2.

The nonperturbative derivation is made possible with an optimal combination of analytic and numeric approaches to solve the invariance equation. The result for the hydrodynamic modes is demonstrated in Fig. 5.2. It is clear from Fig. 5.2 that the relaxation of none of the hydrodynamic modes is faster than $\omega = -1$, which is the collision frequency in the units adopted in this chapter. Thus, the result for the exact hydrodynamics indeed corresponds to the following intuitive picture: the hydrodynamic modes, at large k, cannot relax faster than the (single) collision frequency itself.

We iteratively calculated (i)$\delta\mathbf{X}$ directly from (5.28) for each k in terms of $\mathbf{M}$, and (ii) subsequently calculated moments from $\delta\mathbf{X}$ by either symbolic or numeric integration (both approaches produce the same results within machine precision; we found simple numeric integration on a regular 500×100 grid in c, z-space with grid spacing 0.01 on both axes sufficient to reproduce analytic results. Importantly, the fixed point of the iteration (i)–(ii)–(i)– etc. is unique for each k, i.e., does not depend on the initial values for moments A–Z.

In addition, two other computational strategies were implemented: First, we used continuation of functions A–Z from their values at $k = 0$ to solve (5.28) with an incremental increase of k, where the solution at k was used as the initial guess for $k + \mathrm{d}k$. Second, we also used a "backward" continuation in which the solution at some k (obtained by convergent iterations with a random initial condition) was used as the initial guess for a solution at $k - \mathrm{d}k$. Both these strategies returned the same values of functions A–Z as computed by iterations from arbitrary initial condition.

The solution $\delta \mathbf{X}$ allows us to calculate the whole distribution function f via (5.13), as illustrated in Fig. 5.3. For the resulting moments A–Z, for a wide range of k-values, see Fig. 5.4. With the result for the functions A–Z in hand, the extended hydrodynamic equations are closed.

Let us briefly discuss the pertinent properties of this system. First, the generalized transport coefficients are given by the nontrivial eigenvalues of $-k^{-2}\mathrm{Re}(\mathbf{M})$: $\lambda_2 = -A$ (elongation viscosity), $\lambda_3 = -\frac{2}{3}Y$ (thermal diffusivity), and $\lambda_4 = -D$ (shear viscosity). All these generalized transport coefficients are nonnegative (see Fig. 5.4). Second, in computing the eigenvalues of matrix $\mathbf{M}$, we obtain the dispersion relation $\omega(k)$ of the corresponding hydrodynamic modes already presented in Fig. 5.2. Third, a suitable transform $\mathbf{z}_k = \mathbf{T} \cdot \mathbf{x}_k$ of the hydrodynamic fields, where $\mathbf{T}$ is a real-valued matrix, can be established such that the transformed hydrodynamic equations read $\partial_t \mathbf{z}_k = \mathbf{M}' \cdot \mathbf{z}_k$, where $\mathbf{M}' = \mathbf{T} \cdot \mathbf{M} \cdot \mathbf{T}^{-1}$ is manifestly hyperbolic and stable; $\mathrm{Im}\,(\mathbf{M}')$ is symmetric, and $\mathrm{Re}(\mathbf{M}')$ is symmetric and nonpositive semidefinite. The corresponding transformation matrix $\mathbf{T}$ can be easily read off from the results obtained in [2] for Grad's systems, since the structure of the matrix $\mathbf{M}$ (5.27) is identical to the one studied in [2, 3]. We have explicitly verified that that matrix $\mathbf{T}$ (Eqs. (21)–(23) in

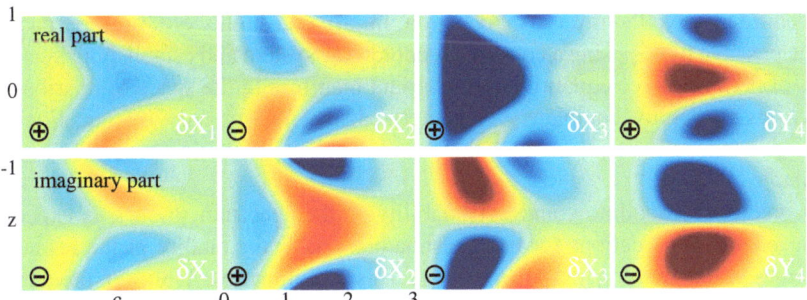

Fig. 5.3 Sample distribution function $f(\mathbf{c}, \mathbf{k})$ at $k = 1$, fully characterized by the four quantities $\delta X_{1,2,3}(c, z)$ and $\delta Y_4(c, z)$. Shown here are both their real (*left column*) and imaginary parts (*right column*). In order to improve contrast, we actually plot $\ln |1 + f^{\mathrm{GM}} \delta X_\mu|$ multiplied by the sign of δX_μ. The same color code for all plots, ranging from -0.2 (*red*) to $+0.2$ (*blue*)

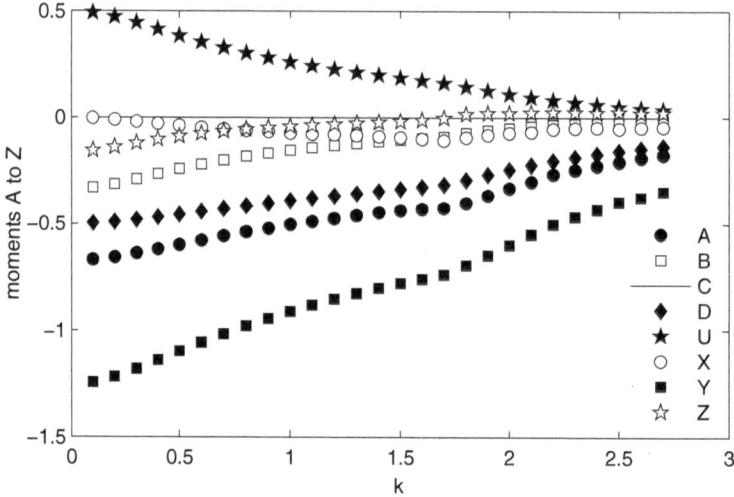

Fig. 5.4 Moments $A–Z$ versus wave number k obtained with the solution of (5.28)

[2] and Eq. (13) in [3]) with the functions $A–Z$ derived herein is real-valued and thus renders the transformed hydrodynamic equations manifestly hyperbolic and stable. We note that this result—hyperbolicity of exact hydrodynamic equations—strongly supports a recent suggestion by Bobylev to consider a hyperbolic regularization of the Burnett approximation [7]. Similarly, using the hyperbolicity, an H-theorem is elementarily proven as in [3, 7]. Finally, using the accurate data for functions $A–Z$, we can write analytic approximations for the corresponding hydrodynamic equations in such a way that hyperbolicity and stability are not destroyed in such an approximation [2].

In conclusion, we have derived exact hydrodynamic equations from the linearized Boltzmann–BGK equation [13]. The main novelty is the numeric nonperturbative procedure to solve the invariance equation. In turn, our highly efficient approach is made possible by choosing a convenient coordinate system and establishing symmetries of the invariance equation. The invariant manifold in the space of distribution functions is thereby completely characterized; that is, not only are equations of hydrodynamics obtained, but also the corresponding distribution function is made available.

The predicted smoothness and extendability of the spectrum to all k is expected to have some implications for microresonators, where the quality of the resonator becomes better at very high frequencies. This is compatible with our prediction. The damping of all the modes saturates, while the imaginary part of the acoustic modes frequency grows. The pertinent data can be used, in particular, as a much needed benchmark for computation-oriented kinetic theories such as lattice Boltzmann models, as well as for constructing novel models [23].

It is worth remarking that the above derivation of hydrodynamics is done under the standard assumption of local equilibrium. However, the assumption itself is open to further study [24].

5.5 The Maxwell Molecules Gas

In this section, we investigate another kinetic model for which an exact solution of the invariance equation (5.19) can be obtained: the Maxwell-molecules gas, i.e., a gas consisting of particles repelling each other with a force proportional to the inverse fifth power of the distance. Chang and Uhlenbeck [25] provided an analytic solution to the eigenvalue problem for the linearized collision operator L pertaining to this case. As anticipated in Chap. 2, in the Maxwell molecules gas, the collision probability per unit time, $g\sigma(g, \theta)$, is independent of the magnitude of the relative velocity g. Since the collision operator is spherically symmetric in the velocity space, the dependence of the eigenfunctions on the direction of $\mathbf{c}$ is expected to be spherically harmonic. Indeed, the eigenvalue problem admits the following solutions:

$$\hat{L}[\psi_{r,l}(c, z)] = \lambda_{r,l}\psi_{r,l}(c, z),$$

$$\psi_{r,l}(c, z) = \sqrt{\frac{r!(l + \frac{1}{2})\sqrt{\pi}}{(l + r + \frac{1}{2})!}} \, c^l P_l(z) S_{l+\frac{1}{2}}^{(r)}(c^2), \tag{5.29}$$

where $S_{l+1/2}^{(r)}(x)$ are Sonine polynomials, and $P_l(z)$ are Legendre polynomials that act on the azimuthal component of the peculiar velocity $\mathbf{c}$. The Legendre and Sonine polynomials are each orthogonal sets, i.e.,

$$\int_{-1}^{1} P_l(z) P_n(z) dz = \frac{2}{2l + 1} \delta_{ln},$$

$$2\pi \int_{0}^{\infty} c^2 e^{-c^2} c^{2l} S_{l+\frac{1}{2}}^{(r)}(c^2) S_{l+\frac{1}{2}}^{(p)}(c^2) dc = \frac{\pi(l + \frac{1}{2} + r)!}{r!} \delta_{rp}.$$

Accordingly, the $\psi_{r,l}$ are normalized to unity with the weight factor $f_0(c)$:

$$\delta_{rr'}\delta_{ll'} = 2\pi^{-1/2} \int_{-1}^{1}\int_{0}^{\infty} c^2 e^{-c^2} \psi_{r,l}(c, z)\psi_{r',l'}(c, z) dc dz$$

$$\equiv \pi^{-3/2} \int e^{-c^2} \psi_{r,l}(\mathbf{c})\psi_{r',l'}(\mathbf{c}) d\mathbf{c}. \tag{5.30}$$

The corresponding eigenvalues for Maxwell molecules are given by

$$\lambda_{r,l} = 2\pi \int_0^\pi \sin(\theta) F(\theta) T_{rl}(\theta) d\theta,$$

$$T_{rl}(\theta) \equiv \cos^{2r+l}\left(\frac{\theta}{2}\right) P_l\left(\cos\frac{\theta}{2}\right) + \sin^{2r+l}\left(\frac{\theta}{2}\right) P_l\left(\sin\frac{\theta}{2}\right) - (1 + \delta_{r0}\delta_{l0}).$$

The collision operator $\hat{L}$ is negative semidefinite, i.e., all eigenvalues are negative except $\lambda_{0,0}$, $\lambda_{0,1}$, and $\lambda_{1,0}$, which are zero and correspond to the elementary collision invariants. As shown by Chang and Uhlenbeck [25], the spectrum of $\hat{L}$ for Maxwell molecules is discrete, and for $r \to \infty$, the eigenvalues $\lambda_{r,l}$ tend to $-\infty$. Chang and Uhlenbeck's investigation of the dispersion of sound in a Maxwell-molecules gas was based on writing the deviation from the global equilibrium as

$$(\varphi_0 + \delta\varphi) = \sum_{\{r,l\}=0}^\infty a^{(r,l)} \psi_{r,l}(\mathbf{c}), \tag{5.31}$$

so that (5.5) reduces to an algebraic equation for the coefficients $a^{(r,l)}$:

$$\omega a^{(r,l)} = -i\mathbf{k} \cdot \sum_{\{r',l'\}=0}^\infty \mathfrak{C}_{r,l,r',l'} a^{(r',l')} + \lambda_{r,l} a^{(r,l)}, \tag{5.32}$$

with

$$\mathfrak{C}_{r,l,r',l'} = \langle \psi_{r,l} | \mathbf{c} \ \psi_{r',l'} \rangle.$$

The hydrodynamic modes for the Maxwell-molecules gas are determined, as seen in Sect. 5.1, from the condition of nontrivial solvability of the linear system (5.32). This approach allows us to solve the eigenvalue problem (5.3) for an arbitrary number of modes, which is made possible just by tuning the number of eigenfunctions taken into account in (5.31).

Another approach [8] is based instead on the expansion of the functions $[\mathbf{X}^{(0)}, \delta\mathbf{X}]$ in terms of the orthonormal basis $\psi_{r,l} = \psi_{r,l}(c,z)$:

$$X_\mu^{(0)}(c,z) = \sum_{r,l}^N a_\mu^{(0)(r,l)} \psi_{r,l}(c,z), \tag{5.33}$$

$$\delta X_\mu(k,c,z) = \sum_{r,l}^N a_\mu^{(r,l)}(k) \psi_{r,l}(c,z). \tag{5.34}$$

The equilibrium coefficients $a_\mu^{(0)}$ are known, and can be determined by taking advantage of the orthogonality of the eigenfunctions as follows:

$$a_\mu^{(0)(r,l)} = \pi^{-\frac{3}{2}} \int e^{-c^2} \psi_{r,l}(c,z) X_\mu^{(0)}(c,z) d\mathbf{c}. \tag{5.35}$$

Inserting (5.33) and (5.34) into the invariance equation (5.19), we obtain the following nonlinear set of algebraic equations for the unknown coefficients $a_\mu^{(r,l)}(k)$:

$$b_\mu^{(r,l)} M_{\mu\nu} = -i\mathbf{k} \cdot \sum_{r',l'}^{N} b_\nu^{(r',l')} \mathfrak{L}_{(r,l,r',l')} + \sum_{r',l'}^{N} a_\nu^{(r',l')} L_{(r,l,r',l')}, \tag{5.36}$$

with $\forall_{\mu,r,l}, b_\mu^{(r,l)} = \left(a_\mu^{(0)(r,l)} + a_\mu^{(r,l)} \right)$, and

$$\mathfrak{L}_{(r,l,r',l')} = \langle \psi_{r,l} | \hat{L} \ \psi_{r',l'} \rangle.$$

For any order of expansion, the solutions of (5.36) characterize an invariant manifold in the phase space. The matrix elements $\mathfrak{L}_{(r,l,r',l')}$ can be easily evaluated in a few kinetic models, such as the linearized BGK model [13] and Maxwell molecules [8]. In particular, the Maxwell-molecules case is recovered by setting

$$\mathfrak{L}_{(r,l,r',l')} = \lambda_{r,l} \delta_{r,r'} \delta_{l,l'},$$

whereas the linearized BGK model is recovered by setting all nonvanishing eigenvalues equal to $\lambda = -1$. The calculation of the coefficients $\mathbf{a}^{(r,l)}$ via the reformulated invariance equation (5.36) is easily achieved. The invariant manifold is fully characterized through these coefficients: the distribution function is determined, and the corresponding matrix $\mathbf{M}$ of linear hydrodynamics as well as moments A–Z are made accessible.

Solving the invariance equation (5.19) and thus obtaining the distribution function via the coefficients $\mathbf{a}^{(r,l)}$, cf. Fig. 5.5, required minor computational effort [8]. The components A–Z of $\mathbf{M}$ are related to the coefficients $\mathbf{a}^{(r,l)}$ by the expressions

$$A = -\frac{ia_2^{(0,2)}}{\sqrt{3}k}, \quad B = -\frac{a_1^{(0,2)}}{\sqrt{3}k^2}, \quad C = -\frac{a_3^{(0,2)}}{\sqrt{3}k^2},$$

$$X = -\frac{i\sqrt{5}a_1^{(1,1)}}{2k}, \quad Y = -\frac{i\sqrt{5}a_3^{(1,1)}}{2k}, \quad Z = -\frac{\sqrt{5}a_2^{(1,1)}}{2k^2},$$

$$D = -\frac{i}{k} \sum_{r,l}^{N} a_4^{(r,l)} \langle c_\| c_\phi | \psi_{r,l} \rangle, \quad U = -\frac{1}{k^2} \sum_{r,l}^{N} a_4^{(r,l)} \left\langle \left(c^2 - \frac{5}{2} \right) c_\perp | \psi_{r,l} \right\rangle. \tag{5.37}$$

In the regime of large Knudsen numbers, the coefficients $\mathbf{a}^{(r,l)}$ may be used, e.g., to directly calculate phoretic accelerations onto moving and rotating convex particles [26], while in the opposite limit of small k, we recover the classical hydrodynamic equations.

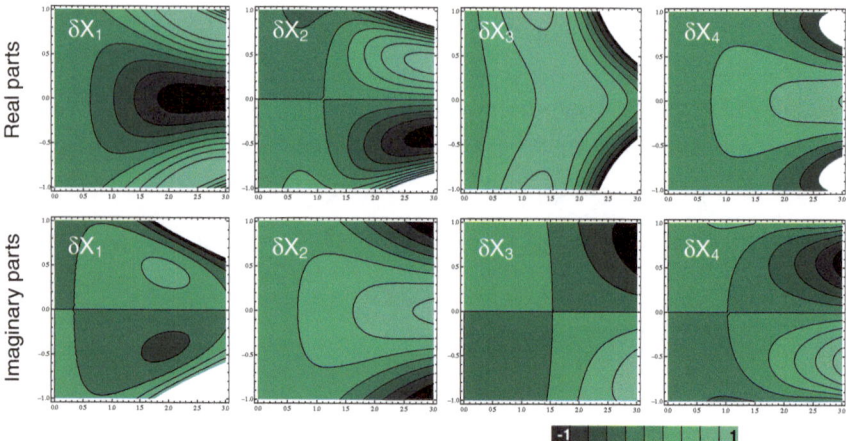

Fig. 5.5 All contributions $\delta X_{1-4}(\mathbf{c}, \mathbf{k})$ versus c (*horizontal*, $c = |\mathbf{c}|$) and $z \in [-1, 1]$ (*vertical axis*, z is the cosine of the angle between $\mathbf{k}$ and peculiar velocity $\mathbf{c}$) to the nonequilibrium distribution function $\delta f = f^{\mathrm{GM}} \delta \mathbf{X} \cdot \mathbf{x}_k$ (5.13) at $k = 1$, obtained with the fourth-order expansion, $N = 4$. Shown here are both their real (*top row*) and imaginary parts (*bottom row*)

5.5.1 Hydrodynamic Modes and Transport Coefficients

With $\mathbf{M}$ in hand, the hydrodynamic modes are obtained from the condition of non-trivial solvability of the linear system (5.17). Figure 5.6 illustrates the damping rates of the fluctuations given by the real part of the hydrodynamic modes, obtained by truncating the series (5.33) and (5.34) at the fourth order. The picture does not quali-tatively change on further increase of the order N. For any given order of expansion, the modes extend smoothly over the entire wave-vector domain, and for large k, they attain an asymptotic value, which is clearly in agreement with the asymptotic behav-ior of the hydrodynamic modes obtained for the linearized BGK model discussed in Sect. 5.4.

The generalized transport coefficients are obtained by the nontrivial eigenvalues of $-k^2 \mathrm{Re}(\mathbf{M})$: $\lambda_2 = -A$ (elongation viscosity), $\lambda_3 = -\frac{2}{3}Y$ (thermal diffusivity), and $\lambda_4 = -D$ (shear viscosity). In the limit $k \to 0$, one recovers the hydrodynamic limit. This limit had been worked out in detail in [2, 3]. In that limit, the generalized transport coefficients A–Z become the classical transport coefficients. As can be seen from Fig. 5.7, and also by inspecting the invariance equation (5.19), in the limit of small k, all moments A–Z approach constant values. These constants are compatible with those obtained in [2, 3] for the case of Navier–Stokes equations and the Burnett correction [27].

The stress tensor and heat flux are given in terms of these moments in Table 5.2. For example, the parallel component of the stress tensor related to density fluctuations, $\sigma_1^{\parallel}$, cf. Eq. (5.25), is given by $-k^2 B$, so that it approaches $-k^2$ for small k, as results from the Burnett approximation [1].

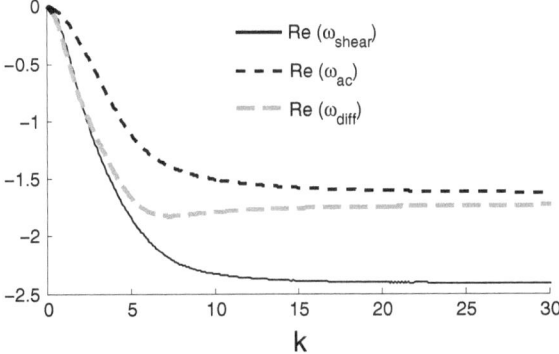

Fig. 5.6 Hydrodynamic modes ω of the Boltzmann kinetic equation with Maxwell-molecules collision operator as a function of wave number k. Shown are two complex conjugate acoustic modes ω_{ac}, twice degenerated shear mode ω_{sh}, and a thermal diffusion mode ω_{diff}

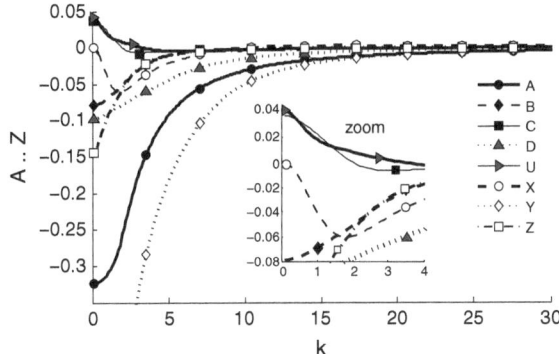

Fig. 5.7 Moments $A - Z$ of the distribution function (see Table 5.2 and Eq. (5.37)) versus wave number k obtained with the solution of (5.19). Non-triangles (*black symbols*): moments entering only the longitudinal component of hydrodynamic equations. Triangles (*blue symbols*): moments entering the transverse component of hydrodynamic equations

Moreover, under suitable assumptions, one may also cast the matrix of hydrodynamic coefficients **M** in the structure of a Green–Kubo formula [10]; cf. Eq. (4.19). We summarize below the main steps of the proof given in [28], to which we refer the reader for an exhaustive derivation. From Eqs. (5.17) and (5.19), the time evolution of the hydrodynamic fields can be formally written as

$$\mathbf{x}_k(\tau) = e^{\mathbf{M}\tau}\mathbf{x}_k(0) = \left\langle \boldsymbol{\xi} \left| e^{\Lambda\tau} \delta \mathbf{X} \right\rangle \mathbf{x}_k(0), \right. \tag{5.38}$$

where we skipped, for brevity, the spatial dependence of the fields. In Eq. (5.38), the time τ is of order τ_{macro}, which denotes a characteristic time scale related to the evolution of the hydrodynamic fields. As discussed in Sect. 3.1, the presence of a definite time scale separation in a particle system entails that $\tau_{macro} \gg \tau_{mf}$, where τ_{mf} denotes a kinetic time scale (i.e., the mean time between collisions). Hence, by invoking the Bogoliubov hypothesis of time scale separation, the time τ becomes large with respect to the characteristic time scale of the dynamics of the distribution function (right-hand side of the second equality in (5.38)). Thus, from Eq. (5.38), we can write the matrix of hydrodynamic coefficients in the form

$$\mathbf{M} = \lim_{\tau \to \infty} \frac{1}{\tau} \log \left\langle \xi \left| e^{\Lambda \tau} \delta \mathbf{X} \right\rangle \right. . \tag{5.39}$$

Next, we use the operator identity

$$e^{\Lambda \tau} = 1 + \Lambda \tau + \Lambda \tau \left[\frac{1}{\tau} \int_0^\tau (\tau - t) e^{\Lambda t} dt \right] \Lambda , \tag{5.40}$$

and make the following assumptions:

(i) The underlying kinetic evolution is such that the term in square brackets in Eq. (5.40) can be approximated, for large τ, by $\int_0^\infty e^{\Lambda t} dt$.
(ii) The expansion of the logarithm to first order in τ is a valid approximation *before* the limit $\tau \to \infty$ is taken.

Thus, using the symmetry of the operator Λ and neglecting, for simplicity, *kinematic contributions* [28] of the form $\langle \xi(0)|\Lambda \delta X(0) \rangle$, Eq. (5.39) can be finally written in the form

$$\mathbf{M} = \int_0^\infty \left\langle \dot{\xi}(0) \delta \dot{\mathbf{X}}(t) \right\rangle_{f^{GM}} dt . \tag{5.41}$$

Equation (5.41) allows us to extend the Green–Kubo formalism, which relates the response function to a suitable time correlation function, to the short-wavelength domain. We should not forget, though, that the standard Green–Kubo response formulas rely on the notion of local thermodynamic equilibrium [29, 30], and hence they are not guaranteed, like the standard relation of equilibrium thermodynamics, below the length scale ℓ_{meso}; cf. Sect. 3.1.

Moreover, it can be seen from an inspection of Fig. 5.7 that in the free-particle regime, $k \to \infty$, the transport coefficients vanish and dissipative effects fade off. This observation finds a sound confirmation in [31]: "Operationally, of course, transport coefficients cannot even be defined for a gas of non-interacting particles. A measurement of the thermal conductivity, for example, is only possible if we can apply, quasi-statically, a temperature gradient and maintain it while we measure the heat current. However, only for a system with a finite mean free path can a temperature gradient be maintained quasi-statically. A free gas would 'run away,' and the standard measurements of transport coefficients cannot be performed. Still, it may

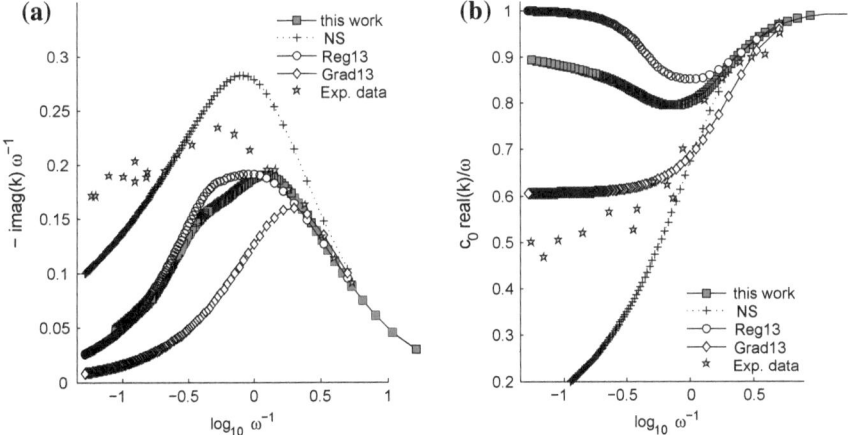

Fig. 5.8 a Damping spectrum, i.e., the negative imaginary part of k divided by frequency ω versus the negative logarithm of ω. Results obtained in this work (by solving Eq. 5.19 and subsequently Eq. 5.18 for $w(k)$ with complex-valued k and real-valued ω) are compared with previous approaches including Navier–Stokes (NS), regularized 13 moment (Reg13) [6], Grad's 13 moment (Grad13), and experimental data presented in [32]. **b** Phase spectrum, i.e., real part of k times velocity of sound c_0 divided by ω versus the negative logarithm of ω. Again, we compare with reference results

be satisfying for some that, in this case, the Kubo expressions give the most sensible result: zero."

5.5.2 Short-Wavelength Hydrodynamics

The existence of collective modes at short wavelengths in real fluids is a longstanding issue in fluid dynamics [33]. In their seminal work [34], Ford et al. illustrated, on the basis of a model kinetic equation approximating the linearized Boltzmann equation, that the sound modes extend to length scales comparable with the mean free path in the gas. Similarly, our analysis showed that hydrodynamic modes and the generalized transport coefficients extend smoothly over the whole k domain. Therefore, the invariant manifold technique allows us to refine the hydrodynamic description beyond the strictly hydrodynamic regime. Our results also strengthen those previously reported in [35–37] on dense fluids, which revealed that the hydrodynamic laws provide a sensible description of fluids even at length scales comparable with λ_{mf}.

It would therefore be interesting to investigate the features of the equations of generalized hydrodynamics that we obtained in the regime of finite frequencies and wave vectors. In Fig. 5.8, a comparison is shown about inverse phase velocity and damping for acoustic waves between our results, former approaches [6, 38], and experimental data performed by Meyer and Sessler [32]. As can be seen, our results are very close to the predictions of the regularized 13 (Reg13) moments method [6] and closer to experimental data than (Reg13) concerning the phase spectrum. Our theory also predicts a phase speed that remains finite also at high frequencies, a property that is not enjoyed by any hydrodynamics derived from the CE expansion [5].

A further clue about the features of our model at finite frequencies and wave vectors can be achieved by investigating the spectrum of density fluctuations $S_{\tilde{n},\tilde{n}}$; cf. Eq. 4.14 in Sect. 4.1. From the knowledge of the functions A–Z, it is possible to compute the coefficients D_T and Γ, related respectively to the damping of thermal and pressure fluctuations in fluids. In the limit of small k, and following standard textbooks [17], they read

$$D_T = \frac{2}{5}(X - Y),$$

$$\Gamma = -\left(\frac{1}{2}A + \frac{1}{5}X + \frac{2}{15}Y\right).$$

It is worth pointing out that in contrast to standard treatments of hydrodynamic fluctuations, the generalized transport coefficient X enters the expression of the coefficients D_T and Γ, even though its contribution, as is evident from Fig. 5.7, is fairly small. The calculation of $S_{\tilde{n},\tilde{n}}$ proceeds along the lines indicated in Chap. 4. We thus give the final result:

$$S_{\tilde{n},\tilde{n}}(k, \omega) = \frac{1}{2\pi} S_{\tilde{n},\tilde{n}}(k) \left[\frac{2}{5}\frac{2D_T k^2}{\omega^2 + (D_T k^2)^2} + \frac{3}{10}\frac{2\Gamma k^2}{(\omega \pm c_0 k)^2 + (\Gamma k^2)^2}\right]. \quad (5.42)$$

Representative plots of $S(k, \omega)$ are shown in Fig. 5.9a, b. For small k (hydrodynamic limit), the obtained spectrum recovers the usual results of neutron (or light) scattering experiments and consists of the three Lorentzian peaks previously illustrated in the left panel of Fig. 4.1. The one centered at $\omega = 0$ is the Rayleigh peak, which corresponds to the diffusive thermal mode. The two side peaks centered at $\omega \pm c_0 k$ are the Brillouin peaks, and represent the two propagating sound waves.

By increasing the wave vector, one enters the regime represented by the red dashed line in the right panel of Fig. 4.1. In this intermediate regime, the structure of (5.42) is unchanged, except that the generalized coefficients D_T and Γ need to be replaced by more complicated expressions [8]. The net effect observed is that sound waves get damped and disappear, whereas the central Rayleigh peak decreases and broadens. Density fluctuations are therefore driven only by a diffusive thermal mode for large enough k. A deeper look into the behavior of the width at half maximum of the central Rayleigh peak with increasing wave vectors allows us to bridge the gap between the hydrodynamic and the free-particle regimes. As discussed in Sect. 4.2, for $k \ll 1$, the width of the central peak increases with the square wave vector $\propto k^2$, whereas in the opposite regime, $k \gg 1$, the width of the central peak is expected to grow linearly in k.

Our results, see Fig. 5.9c, predict a width that is truly quadratic for small enough k, reaches the regime of linear behavior for large k, and terminates, for some large k, with a sublinear dependence on k. The onset of the terminal regime at $k = k^*(N)$ marks the range of validity that can be accessed at a given finite order of expansion

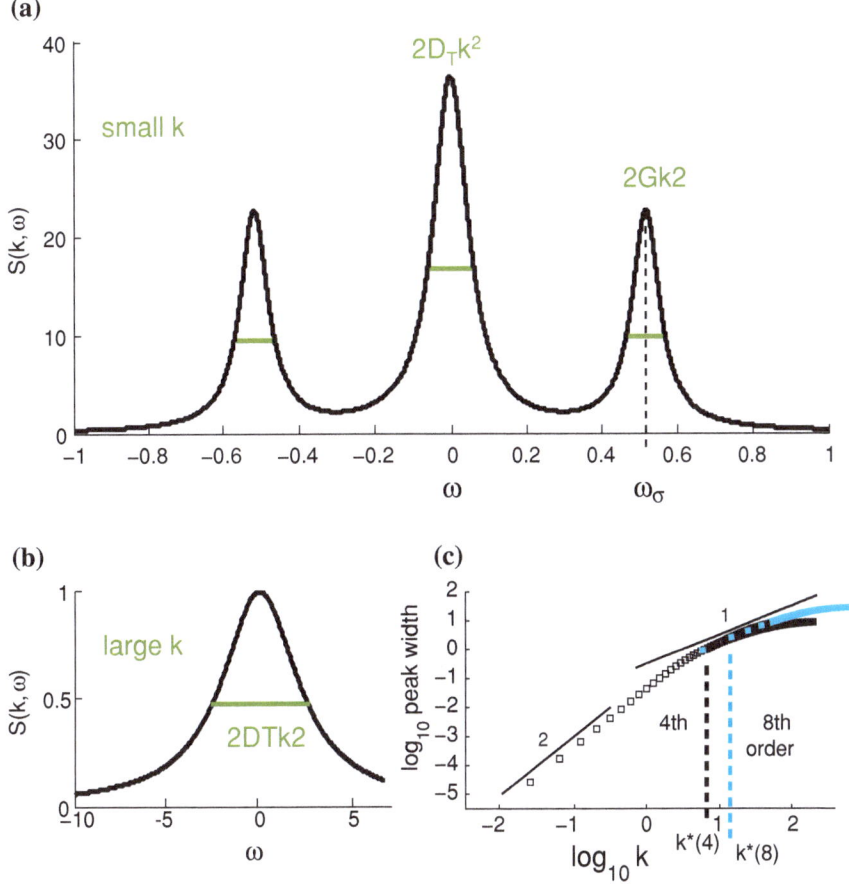

Fig. 5.9 **a** Dynamic structure factor $S_{\tilde{n},\tilde{n}}(k, \omega)$ versus ω for small $k = 0.4$ and **b** large $k = 100$. $\omega_s = c_0 k$ denotes the hydrodynamic predicted sound mode of the spectrum, and the widths are related to the moments A–Z (see Fig. 5.7). For small k, these are given by $D_T = \frac{2}{3}(X - Y)$ and $\Gamma = -(\frac{1}{2}A + \frac{1}{5}X + \frac{2}{15}Y)$, where A is the generalized longitudinal kinetic viscosity, Y is the generalized thermal diffusion coefficient, and X is a cross-coupling transport coefficient, relating heat flux to density gradients. **c** Width $D_T k^2$ of the Rayleigh peak versus k (double-logarithmic). At small k, $D_T k^2 \propto k^2$ as all moments A–Z, except X, reach a finite value in this limit. The inflection point at $k = k^*(N) \gg 1$ (shown to be increasing with the order of expansion N) denotes the onset of departure from the ideal Maxwellian behavior, where the width of the peak starts to behave sublinearly in k and is used to quantify the range of validity for results obtained at finite order

N. Increasing N does not alter the overall picture obtained at a moderate order of expansion, and more generally, results obtained with $N + 1$ will not change those obtained with N below $k^*(N)$; cf. Fig. 5.9c.

In the spirit of Grad's moments technique, cf. Sect. 3.3, varying the parameter N in the expansions (5.33) and (5.34) corresponds to tuning the number of nonequilibrium contributions to be included in the definition of distribution function. This leads to

an arbitrary level of refinement in the accuracy of the hydrodynamic description and explains, intuitively, the observed dependence of k^* on N [8].
The onset of a critical length scale marking the limit of the hydrodynamic description is also reminiscent of the discussion in Chap. 3. Namely, we know that hydrodynamics is founded on the notion of local thermodynamic equilibrium. Thus, for a given N, one may regard the length scale at which hydrodynamics breaks down, $\left[k^*(N)\right]^{-1}$, as inherently related to the scale ℓ_{meso} below which local equilibrium is lost. Further investigation is called for to shed light on this connection, which, if confirmed, would endow the generalized hydrodynamic equations obtained via invariant manifold theory with an even stronger thermodynamic character.

References

1. A.N. Gorban and I.V. Karlin, *Invariant Manifolds for Physical and Chemical Kinetics*, Lect. Notes Phys. **660** (Springer, Berlin, 2005).
2. M. Colangeli, I.V. Karlin and M. Kröger, From hyperbolic regularization to exact hydrodynamics for linearized Grad's equations, *Phys. Rev. E* **75**, 051204 (2007).
3. M. Colangeli, I.V. Karlin and M. Kröger, Hyperbolicity of exact hydrodynamics for three-dimensional linearized Grad's equations, *Phys. Rev. E* **76**, 022201 (2007).
4. A.V. Bobylev, *Sov. Phys. Dokl.* **27**, 29 (1982).
5. H. Struchtrup, *Macroscopic Transport Equations for Rarefied Gas Flows. Approximation Methods in Kinetic Theory*, (Springer-Verlag Berlin Heidelberg 2005).
6. H. Struchtrup and M. Torrilhon, H-theorem, regularization, and boundary conditions for linearized 13 moment equations, *Phys. Rev. Lett.* **99**, 014502 (2007).
7. A.V. Bobylev, Instabilities in the Chapman-Enskog expansion and hyperbolic Burnett equations, *J. Stat. Phys.* **124**, 371 (2006).
8. M. Colangeli, M. Kröger and H. C.Öttinger, Boltzmann equation and hydrodynamic fluctuations, *Phys. Rev. E* **80**, 051202 (2009).
9. P.L. Bhatnagar, E.P. Gross and M. Krook, A Model for Collision Processes in Gases. I. Small Amplitude Processes in Charged and Neutral One-Component Systems, *Phys. Rev.* **94**, 511 (1954).
10. U. Marini Bettolo Marconi, A. Puglisi, L. Rondoni and A. Vulpiani, Fluctuation-Dissipation: Response Theory in Statistical Physics, *Phys. Rep.* **461**, 111 (2008).
11. L. Onsager and S. Machlup, Fluctuations and irreversible processes, *Phys. Rev.* **91**, 1505 (1953).
12. J. L. Lebowitz, E. Presutti and H. Spohn, Microscopic Models of Hydrodynamic Behavior, *J. Stat. Phys.* **51**, 841 (1988).
13. I.V. Karlin, M. Colangeli and M. Kröger, Exact linear hydrodynamics from the Boltzmann Equation, *Phys. Rev. Lett.* **100**, 214503 (2008).
14. P. Resibois, On linearized hydrodynamic modes in statistical physics, *J. Stat. Phys.* **2**, 1 (1970).
15. J. M. Blatt, Model equations in the kinetic theory of gases, *J. Phys. A: Math. Theor.* **8**, 980 (1975).
16. R. Balescu, *Equilibrium and nonequilibrium statistical mechanics* (Wiley, 1975).
17. L. E. Reichl, *A modern course in statistical physics* (University of Texas Press, Austin, 1980).
18. M. Kröger, *Models for Polymeric and Anisotropic Liquids* (Springer, Berlin, 2005).
19. S. Hess and W. Köhler, *Formeln zur Tensor-Rechnung* (Palm & Enke, Erlangen, 1980).
20. M. Abramowitz and I.A. Stegun, *Handbook of mathematical functions* (National Bureau of Standards, Washington, 1967).
21. J.P. Boon and S. Yip, *Molecular Hydrodynamics* (Dover, 1991).

22. C. Cercignani, *Theory and Application of the Boltzmann Equation* (Scottish Academic Press, Edinburgh, 1975).
23. S.S. Chikatamarla, S. Ansumali and I.V. Karlin, Entropic Lattice Boltzmann Models for Hydrodynamics in Three Dimensions, *Phys. Rev. Lett.* **97**, 010201 (2006).
24. H. C.Öttinger and H. Struchtrup, The mathematical procedure of coarse graining: From Grad's ten-moment equations to hydrodynamics, *Multiscale Model. Simul.* **6**, 53 (2007).
25. C. S. Wang Chang and G.E. Uhlenbeck, *The kinetic theory of gases* (Amsterdam, North-Holland Pub. Co., 1970).
26. M. Kröger and M. Hütter, Unifying kinetic approach to phoretic forces and torques for moving and rotating convex particles, *J. Chem. Phys.* **125**, 044105 (2006).
27. D. Burnett, The distribution of velocities and mean motion in a slight nonuniform gas, *Proc. London Math. Soc.* **39**, 385 (1935).
28. A. N. Gorban and I. V. Karlin, Reconstruction lemma and fluctuation-dissipation theorem, *Revista Mexicana de Fisica* **48**, Suppl. 1, 238 (2002).
29. L. Rondoni and E. G. D. Cohen, On some derivations of irreversible thermodynamics from dynamical systems theory, *Physica D* **341** 168, (2002).
30. M. Colangeli and L. Rondoni, Equilibrium, fluctuation relations and transport for irreversible deterministic dynamics, *Physica D* **241** 681 (2012).
31. D. Forster, *Hydrodynamic fluctuations, Broken Symmetry, and Correlation Functions* (W. A. Benjamin, New York, 1975).
32. E. Meyer and G. Sessler, Schallausbreitung in Gasen bei hohen Frequenzen und sehr niedrigen Drucken, *Z. Phys.* **149**, 15 (1947).
33. I. M. de Schepper and E. G. D. Cohen, Very-short-wavelength collective modes in fluids, *J. Stat. Phys.* **27**, 2 (1982).
34. J. D. Foch and G. W. Ford, in Studies in Statistical Mechanics (North-Holland, Amsterdam, 1970).
35. B.J. Alder and W.E. Alley, Generalized hydrodynamics, *Phys. Today* **37**, 56 (1984).
36. B. J. Alder and W.E. Alley, Generalized transport coefficients for hard spheres, *Phys. Rev. A* **27**, 3158 (1983).
37. T. R. Kirkpatrick, Short-wavelength collective modes and generalized hydrodynamic equations for hard-sphere particles, *Phys. Rev. A* **32**, 3130 (1985).
38. H. Grad, On the kinetic theory of rarefied gases, *Comm. Pure and Appl. Math.* **2**, 331 (1949).

Chapter 6
Grad's 13-Moments System

In this chapter, we will describe a procedure that allows us to reduce the description from Grad's moment system [1] to the hydrodynamic level.

We will also provide a comparison of our results with the standard CE procedure, introduced in Sect. 3.2, which is based on a formal expansion of the stress tensor and heat flux vector in terms of derivatives of the hydrodynamic fields. Truncating the expansion to the first power of the Knudsen number yields the NSF equations, while next-order approximations lead to the so-called Burnett [2] (ε^2) and super-Burnett (ε^3) corrections.

It has long been conjectured that the inclusion of higher-order terms in the constitutive relations for the stress tensor and heat flux may improve the predictive capabilities of hydrodynamics formulations in the continuum–transition regime where NSF equations fail. However, Bobylev's investigation of Maxwell molecules [3] proved that the Burnett and the super-Burnett hydrodynamics may violate the basic physics behind the Boltzmann equation. Namely, sufficiently short acoustic waves increase with time instead of decaying. Bobylev's instability has been also studied by Uribe et al. [4] for hard-sphere molecules. This instability contradicts the H-theorem, since all near-equilibrium perturbations are expected to decay, and it creates difficulties for an extension of hydrodynamics into a highly nonequilibrium domain where the NSF approximation is inapplicable. For example, higher-order systems of hydrodynamic equations afforded a better description in certain situations such as shock structures on coarse grids, but are prone to small-wavelength instabilities when grids are refined.

Successes and drawbacks of the Burnett computations and a family of extended Burnett equations aimed at achieving entropy-consistent behavior of the equations have recently been summarized in [5]. As discussed in [6], the failure of the CE expansion is a consequence of the truncation of the expansion. This question was studied in some detail for a class of simple kinetic models—Grad's moment systems—in [6–11]. In these works, the CE expansion was summed up exactly, which revealed the stability of the exact hydrodynamics, in contrast to its finite-order approximations.

M. Colangeli, *From Kinetic Models to Hydrodynamics*, SpringerBriefs in Mathematics, DOI: 10.1007/978-1-4614-6306-1_6, © Matteo Colangeli 2013

Alternative ways of approximating the CE solution have been also suggested. Very recently, Bobylev suggested a different viewpoint on the problem of Burnett's hydrodynamics [12]. Namely, violation of hyperbolicity can be seen as a source of instability. We recall that Boltzmann's and Grad's equations are hyperbolic and stable due to corresponding H-theorems. However, the Burnett hydrodynamics is not hyperbolic, which leads to no H-theorem. Bobylev [12] suggested that one stipulate hyperbolization of Burnett's equations, which can also be considered a change of variables. In this way, hyperbolically regularized Burnett's equations admit the H-theorem (in the linear case, at least), and stability is restored.

Thus, we aim at studying the issue of hyperbolicity of higher-order hydrodynamics in the case that the CE solution can be found exactly.

This chapter is organized as follows: In Sect. 6.1, we will briefly review the derivation of the 13-moment system from the Boltzmann equation. In Sect. 6.2, we will then discuss the derivation of hydrodynamics from Grad's moment system, linearized around equilibrium and assuming unidirectional flow conditions (i.e., the so-called 1D13M system, according to the notation introduced in [6, 13]). While simple enough, this model is nontrivial for three reasons:

- Application of the CE method leads to rather involved nonlinear recurrence relations for the coefficients of the expansion.
- The Burnett approximation derived from Grad's moment system is identical to the one derived from the Boltzmann equation for Maxwell molecules and thus violates hyperbolicity and exhibits Bobylev's instability [3, 4].
- Even though the exact hydrodynamics can be derived following the lines of [6–11] and is stable, the question remains whether this exact hydrodynamics is manifestly hyperbolic.

We will also comment on the properties of the obtained hydrodynamic solutions. In particular, we will focus on the onset of a critical value of the Knudsen number beyond which the hydrodynamic description breaks down. Finally, the three-dimensional extension of the model will be illustrated in Sect. 6.3.

6.1 Derivation of the 13-Moment System from the Boltzmann Equation

Let us briefly sketch here the derivation of Grad's 13-moment system from the Boltzmann equation. We consider a set of distinguished fields defined as follows:

$$\mathbf{x}_G = \mathbf{x} + \mathbf{x}_1.$$

The first term, $\mathbf{x} \equiv [\tilde{n}, \tilde{\mathbf{u}}, \tilde{T}]$, denotes the set of dimensionless hydrodynamic fluctuations introduced in Sect. 5.2. The second term is defined as $\mathbf{x}_1 \equiv [\tilde{\sigma}, \tilde{\mathbf{q}}]$, where $\tilde{\sigma} = \delta\sigma/p_0$ and $\tilde{\mathbf{q}} = \mathbf{q}/(p_0 v_T)$ denote respectively the dimensionless stress tensor and heat flux. Here $p_0 = n_0 k_B T_0$ is the equilibrium hydrostatic pressure. Let $\mathbf{X}^{(0)}$

and $\boldsymbol{\xi}^{(0)}$ be the vectors expressed by Eqs. (5.21)–(5.22). We also introduce the vectors $\mathbf{X}^{(1)}$ and $\boldsymbol{\xi}^{(1)}$, defined as

$$\mathbf{X}^{(1)}(\mathbf{c}) = \left(\overline{\mathbf{cc}}, \frac{4}{5}\mathbf{c} \left(c^2 - \frac{5}{2} \right) \right),$$

$$\boldsymbol{\xi}^{(1)}(\mathbf{c}) = \left(2\,\overline{\mathbf{cc}}, \mathbf{c} \left(c^2 - \frac{5}{2} \right) \right).$$

The linearized Grad's 13-moment distribution function f_G can be written in the form

$$f_G(\mathbf{r}, \mathbf{c}, t) = f^{GM}(1 + \varphi_G) = f^{GM}(1 + \varphi + \varphi_1) = f^{GM} \left(1 + \mathbf{X}^{(0)} \cdot \mathbf{x} + \mathbf{X}^{(1)} \cdot \mathbf{x}_1 \right); \tag{6.1}$$

cf., for analogy, Eq. (5.24). One can immediately verify that the nonhydrodynamic fields are obtained via

$$\mathbf{x}_1 = \left\langle \boldsymbol{\xi}^{(1)}(\mathbf{c}) \right\rangle_{f_G}. \tag{6.2}$$

Inserting the ansatz (6.1) into the linearized Boltzmann equation (5.3), one obtains

$$\partial_t \varphi_G = \Lambda \varphi_G,$$

where Λ is the linear operator defined in Sect. 5.1. Next, we introduce the projection operator $P_G = P + P_1$, where P is defined as in Eq. (5.15), whereas P_1 reads

$$P_1 \Lambda \varphi_G = D_{\mathbf{x}_1} \varphi_G \cdot \int \boldsymbol{\xi}^{(1)}(\mathbf{c}) \Lambda \varphi_G \mathrm{d}\mathbf{v}.$$

The resulting equation,

$$P_G[f^{GM} \partial_t \varphi_G] = P_G[f^{GM} \Lambda \varphi_G], \tag{6.3}$$

expresses, in a compact form, Grad's 13-moment system for the variables $\mathbf{x}_G = [\tilde{n}, \tilde{\mathbf{u}}, \tilde{T}, \tilde{\sigma}, \tilde{\mathbf{q}}]$.

6.2 Hydrodynamics from the Linearized One-Dimensional Grad's System

Equation (6.3) corresponds to the following set of equations in one spatial variable x:

$$\partial_t \tilde{\rho} = -\partial_x \tilde{u},$$

$$\partial_t \tilde{u} = -\partial_x \tilde{\rho} - \partial_x \tilde{T} - \partial_x \tilde{\sigma},$$

$$\partial_t \tilde{T} = -\frac{2}{3}\partial_x \tilde{u} - \frac{2}{3}\partial_x \tilde{q},$$

$$\partial_t \tilde{\sigma} = -\frac{4}{3}\partial_x \tilde{u} - \frac{8}{15}\partial_x \tilde{q} - \frac{1}{\varepsilon}\tilde{\sigma},$$

$$\partial_t \tilde{q} = -\frac{5}{2}\partial_x \tilde{T} - \partial_x \tilde{\sigma} - \frac{2}{3\varepsilon}\tilde{q}. \tag{6.4}$$

Here, $\tilde{\rho}(x,t) = \tilde{n}(x,t)m$, $\tilde{u}(x,t)$, and $\tilde{T}(x,t)$ denote the reduced deviations of mass density, average velocity, and temperature from their equilibrium values, whereas $\tilde{\sigma}(x,t)$ and $\tilde{q}(x,t)$ are reduced xx-components of the nonequilibrium stress tensor and heat flux, respectively. The parameter $\varepsilon > 0$ denotes, as usual, the Knudsen number. The system (6.4) provides the time evolution equations for the set $\mathbf{x}$ of hydrodynamic (locally conserved) fields coupled to the set $\mathbf{x}_1$ of nonhydrodynamic fields. The goal is to reduce the number of equations in (6.4) and to arrive at a closed system of three equations for the hydrodynamic fields only.

Thanks to the linearity of the system (6.4), it proves convenient to look at the reciprocal space and seek solutions of the form $\zeta = \zeta_k \exp(\omega t + ikx)$, where ζ is a generic function $[\rho, u^{\parallel}, T, \sigma^{\parallel}, q^{\parallel}]$. In the sequel, we use rescaled variables $t' = \varepsilon^{-1}t$, $x' = \varepsilon^{-1}x$, and $k' = \varepsilon k$, cf. Eq. (2.56), and omit the prime and tilde to simplify notation. Application of the CE method, as illustrated in Sect. 3.2, to the reduction of the system (6.4) results in the following series expansion of the nonhydrodynamic variables:

$$\sigma_k = \sum_{n=0}^{\infty} \sigma_k^{(n)}, \quad q_k = \sum_{n=0}^{\infty} q_k^{(n)}, \tag{6.5}$$

where the coefficients $\sigma_k^{(n)}$ and $q_k^{(n)}$ are obtained from a recurrence procedure,

$$\sigma_k^{(n)} = -\left\{ \sum_{m=0}^{n-1} \partial_t^{(m)} \sigma_k^{(n-1-m)} + \frac{8}{15}ikq_k^{(n-1)} \right\},$$

$$q_k^{(n)} = -\left\{ \sum_{m=0}^{n-1} \partial_t^{(m)} q_k^{(n-1-m)} + ik\sigma_k^{(n-1)} \right\}, \tag{6.6}$$

and where the CE operators $\partial_t^{(m)}$ act on the hydrodynamic fields as follows:

$$\partial_t^{(m)} \rho_k = \begin{cases} -iku_k & , m = 0 \\ 0 & , m \geq 1 \end{cases},$$

$$\partial_t^{(m)} u_k = \begin{cases} -ik(\rho_k + T_k) & , m = 0 \\ -ik\sigma_k^{(m-1)} & , m \geq 1 \end{cases},$$

$$\partial_t^{(m)} T_k = \begin{cases} -\frac{2}{3}iku_k & , m = 0 \\ -\frac{2}{3}ikq_k^{(m-1)} & , m \geq 1 \end{cases}. \tag{6.7}$$

It can be proven that the functions σ_k and q_k have the following structure for all $n = 0, 1, \ldots$:

$$\sigma_k^{(2n)} = a_n(-k^2)^n iku_k,$$
$$\sigma_k^{(2n+1)} = b_n(-k^2)^{n+1}\rho_k + c_n(-k^2)^{n+1}T_k,$$
$$q_k^{(2n)} = x_n(-k^2)^n ik\rho_k + y_n(-k^2)^n ikT_k,$$
$$q_k^{(2n+1)} = z_n(-k^2)^{n+1}u_k, \tag{6.8}$$

where $a_n, \ldots, z_n$ are numerical coefficients to be determined. Note the alternating structure of expansion coefficients of odd and even orders. Substituting (6.8) into ..., the CE method is cast into the form of recurrence equations in terms of the coefficients $a_n, \ldots, z_n$:

$$a_{n+1} = b_n + \frac{2}{3}c_n + \frac{2}{3}\sum_{m=1}^{n} c_{n-m}z_{m-1} + \sum_{m=0}^{n} a_{n-m}a_m - \frac{8}{15}z_n,$$

$$b_{n+1} = a_{n+1} + \sum_{m=0}^{n} a_{n-m}b_m + \frac{2}{3}\sum_{m=0}^{n} c_{n-m}x_m - \frac{8}{15}x_{n+1},$$

$$c_{n+1} = a_{n+1} + \sum_{m=0}^{n} a_{n-m}c_m + \frac{2}{3}\sum_{m=0}^{n} c_{n-m}y_m - \frac{8}{15}y_{n+1},$$

$$x_{n+1} = z_n + \sum_{m=1}^{n} z_{n-m}b_{m-1} + \frac{2}{3}\sum_{m=0}^{n} y_{n-m}x_m - b_n,$$

$$y_{n+1} = z_n + \sum_{m=1}^{n} z_{n-m}c_{m-1} + \frac{2}{3}\sum_{m=0}^{n} y_{n-m}y_m - c_n,$$

$$z_{n+1} = x_{n+1} + \frac{2}{3}y_{n+1} + \frac{2}{3}\sum_{m=0}^{n} y_{n-m}z_m + \sum_{m=0}^{n} z_{n-m}a_m - a_{n+1}. \tag{6.9}$$

System (6.9) is solved recursively subject to the initial conditions

$$a_0 = -\frac{4}{3}, \ b_0 = -\frac{4}{3}, \ c_0 = \frac{2}{3}, \ x_0 = 0, \ y_0 = -\frac{15}{4}, \ z_0 = -\frac{7}{4}. \tag{6.10}$$

The initial conditions are obtained by evaluating the functions σ_k and q_k up to the Burnett order (see Eq. (6.18) below) and identifying the coefficients a_0, x_0, and y_0 from the Navier–Stokes approximation and the remaining coefficients b_0, c_0, and z_0 from the Burnett correction. Equation (6.9) defines six functions,

$$A(k) = \sum_{n=0}^{\infty} a_n(-k^2)^n, \ldots, Z(k) = \sum_{n=0}^{\infty} z_n(-k^2)^n. \tag{6.11}$$

Thus, the CE solution amounts to finding functions $A, \ldots, Z$ (6.11). Note that by the nature of the CE recurrence procedure, functions $A, \ldots, Z$ (6.11) are real-valued functions. Knowing $A, \ldots, Z$ (6.11), we can express the nonequilibrium stress tensor and heat flux as

$$\sigma_k = ikA(k)u_k - k^2 B(k)\rho_k - k^2 C(k)T_k, \qquad (6.12)$$

$$q_k = ikX(k)\rho_k + ikY(k)T_k - k^2 Z(k)u_k. \qquad (6.13)$$

On substituting these expressions into the Fourier-transformed balance equation (6.4), we obtain a closed system of hydrodynamic equations, which is conveniently written in vector form,

$$\partial_t \mathbf{x}_k = \mathbf{M}\mathbf{x}_k, \qquad (6.14)$$

where $\mathbf{x}_k \equiv (\rho_k, u_k, T_k)$, and the matrix $\mathbf{M}$ has the form

$$\mathbf{M} = \begin{pmatrix} 0 & -ik & 0 \\ -ik(1-k^2 B) & k^2 A & -ik(1-k^2 C) \\ \frac{2}{3}k^2 X & -\frac{2}{3}ik(1-k^2 Z) & \frac{2}{3}k^2 Y \end{pmatrix}. \qquad (6.15)$$

It is worth noticing that the matrix (6.15) obtained here from Grad's moment system (6.4) enjoys the same formal structure of the matrix (5.27) derived from the Boltzmann equation. With $\mathbf{M}$, we find the dispersion relation for the hydrodynamic modes $\omega(k)$ by solving the characteristic equation

$$\det(\mathbf{M} - \omega\mathbf{I}) = 0, \qquad (6.16)$$

with $\mathbf{I}$ the unit matrix. The standard application of the CE procedure is to approximate functions $A, \ldots, Z$ by polynomials with coefficients found from the recurrence procedure (6.9). The first nonvanishing contribution is the Newton–Fourier constitutive relations

$$\sigma_k^{(0)} = -\frac{4}{3}iku_k, \; q_k^{(0)} = -\frac{15}{4}ikT_k, \qquad (6.17)$$

which lead to the NSF hydrodynamic equations. Computing the coefficients $\sigma_k^{(1)}$ and $q_k^{(1)}$, we arrive at the Burnett level:

$$\sigma_k = -\frac{4}{3}iku_k + \frac{4}{3}k^2\rho_k - \frac{2}{3}k^2 T_k,$$

$$\mathbf{q}_k = -\frac{15}{4}ikT_k + \frac{7}{4}k^2 u_k. \qquad (6.18)$$

The Burnett approximation (6.18) coincides with that obtained by Bobylev [3] from the Boltzmann equation for Maxwell molecules. Unlike the NSF approximation,

Fig. 6.1 Dispersion relation. Acoustic mode $\mathrm{Re}(\omega_{ac})$ for NSF and Burnett hydrodynamics

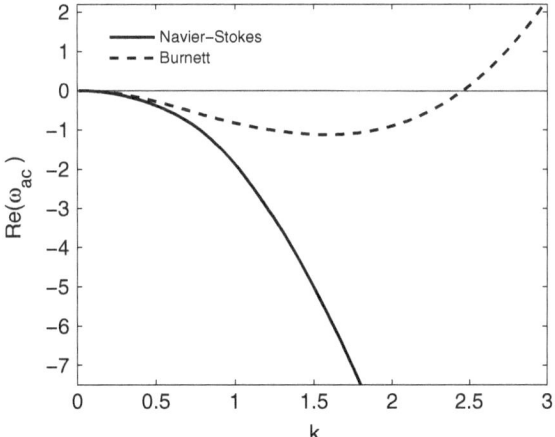

the Burnett constitutive relations (6.18) show instability of the acoustic mode; see Fig. 6.1.

Thus, the difficulty of the CE method consists in the way the functions $A, \ldots, Z$ are approximated. The standard polynomial approximations lead to unstable hydrodynamic equations. We shall now derive closed-form equations for these functions, which amounts to summing the CE series exactly.

6.2.1 Invariance Equations

Summation of the CE series for the functions $A, \ldots, Z$ can be done directly from the recurrence relations (6.9) following the approach outlined in [9]. Alternatively, one can solve the invariance equation relevant to Grad's system (6.3). Here, the set of non-hydrodynamic moments $\mathbf{x}_{1,k} \equiv \{\sigma_k, q_k\}$ is still thought of as being in the form (6.12) and (6.13), but the method makes no assumption about the power-series representation of the functions $A, \ldots, Z$. The time derivative of $\{\sigma_k, q_k\}$ can be computed in two different ways. On the one hand, substituting (6.12) and (6.13) into the moment system (6.4), we obtain

$$\partial_t \sigma_k = -\frac{4}{3} i k u_k - \frac{8}{15} i k q(X, Y, Z, k) - \sigma_k(A, B, C, k),$$

$$\partial_t q_k = -\frac{5}{2} i k T_k - i k \sigma_k(A, B, C, k) - \frac{2}{3} q(X, Y, Z, k).$$

On the other hand, the time derivative of $\{\sigma, q\}$ can be computed due to the closed hydrodynamic equations by chain rule:

$$\partial_t \sigma_k = \frac{\partial \sigma_k}{\partial u_k} \partial_t u_k + \frac{\partial \sigma_k}{\partial \rho_k} \partial_t \rho_k + \frac{\partial \sigma_k}{\partial T_k} \partial_t T_k,$$

$$\partial_t q_k = \frac{\partial q_k}{\partial u_k} \partial_t u_k + \frac{\partial q_k}{\partial \rho_k} \partial_t \rho_k + \frac{\partial q_k}{\partial T_k} \partial_t T_k.$$

Here, the derivatives $\partial_t u_k$ and $\partial_t T_k$ are evaluated self-consistently using the functions (6.12) and (6.13) on the right-hand side of (6.4). The two time derivatives coincide, since the set $\{\sigma_k, q_k\}$ has to solve both the full Grad's system and the reduced system. We thereby obtain the invariance equation corresponding to Eq. (6.3) relating the six functions $A(k), \ldots, Z(k)$:

$$-\frac{4}{3} - A - k^2 \left(A^2 + B - \frac{8Z}{15} + \frac{2C}{3} \right) + \frac{2}{3} k^4 CZ = 0,$$

$$\frac{8}{15} X + B - A + k^2 AB + \frac{2}{3} k^2 CX = 0,$$

$$\frac{8}{15} Y + C - A + k^2 AC + \frac{2}{3} k^2 CY = 0,$$

$$A + \frac{2}{3} Z + k^2 ZA - X - \frac{2}{3} Y + \frac{2}{3} k^2 YZ = 0,$$

$$k^2 B - \frac{2}{3} X - k^2 Z + k^4 ZB - \frac{2}{3} YX = 0,$$

$$-\frac{5}{2} - \frac{2}{3} Y + k^2 (C - Z) + k^4 ZC - \frac{2}{3} k^2 Y^2 = 0. \qquad (6.19)$$

The same equations can be derived by summation of the CE expansion. Equations (6.19) are a convenient starting point for evaluation of exact hydrodynamics. For $k = 0$, one recovers the initial conditions (6.10).

6.2.2 Exact Hydrodynamic Solutions

The dispersion relation $\omega(k)$ was found by simultaneously solving numerically (6.19) and the characteristic equation (6.16).

The resulting hydrodynamic spectrum consist of two modes: the acoustic mode $\omega_{ac}(k)$, represented by two complex-conjugate roots of (6.16), and the real-valued diffusive heat mode $\omega_{diff}(k)$; cf. Fig. 6.2. Among the many sets of solutions $\{A(k), \ldots, Z(k)\}$ to the system (6.19), the relevant ones are continuous functions with the asymptotics $\lim_{k \to 0} \omega = 0$. Remarkably, we find that the solution with these asymptotics is unique and is represented by a pair of complex-conjugate sets $\{S, S^*\}$, shown in Figs. 6.3 and 6.4.

Note that a qualitative change of dynamics arises when the diffusive mode couples with one of the two nonhydrodynamic modes of Grad's system at a critical wave number $k_c \approx 0.3023$ (which is also the value where the Newton method diverges;

see below). From the CE perspective, the hydrodynamics of the diffusive mode stops at k_c, since after that point, it becomes a complex-valued function coupled with the conjugate nonhydrodynamic mode; see Fig. 6.2. Essentially, for $k \geq k_c$, the CE method no longer recognizes the resulting diffusive branch as an extension of a hydrodynamic branch. Also, the set of solutions $\{S, S^*\}$, real-valued for $k \leq k_c$, continues on a complex manifold; cf. Fig. 6.3.

This feature has no correspondence with the behavior of the hydrodynamic modes obtained in Chap. 5 from the Boltzmann equation and originates from the lack of dynamical invariance of the underlying Grad's system.

We observe that the occurrence of a pair of complex-conjugate sets of solutions is very plausible due to symmetry: inserting S into the dispersion relation, we obtain a pair of complex-conjugate acoustic modes $[\omega_{ac}(S, k), \omega_{ac}^*(S, k)]$ plus one of the complex modes resulting from the extension of the diffusive branch for $k \geq k_c$, whereas through S^*, we obtain, symmetrically, the two latter conjugate modes, plus one of the conjugate acoustic modes.

As further evidence of this close coupling, we also observe the occurrence of an intersection between the real parts of the hydrodynamic modes $\text{Re}(\omega_{ac})$ and $\text{Re}(\omega_{\text{diff}})$ after the critical point, at $k = k_{\text{coupl}} \approx 0.383$. Therefore, the message extracted from the study of Grad's system (6.4) is that the set of hydrodynamic equations for $[\rho, u, T]$ provides, as expected, stable solutions when one takes into account all the orders of CE expansion—which corresponds to solving the system of invariance equations (6.19).

Moreover, the other relevant observation is that there is no closed set of hydrodynamic equations after k_c, even though the acoustic mode extends smoothly beyond k_c, as is visible in Fig. 6.2. The presence of a coupling at a critical wave vector k_c between hydrodynamic and nonhydrodynamic modes can correspond to the loss of a definite time scale separation, for $k > k_c$, between *slow* and *fast* moments, which, in the Bogoliubov theory outlined in Chap. 3, was one of the basic assumptions needed to interpret the onset of collective behavior in a many-particle system. Thus, the exact hydrodynamics as derived by the summation of the CE expansion (or equivalently, from the invariance equations) extends up to a finite critical value k_c. No stability violation occurs, in contrast to the finite-order truncations thereof [14, 15]. Furthermore, from an inspection of Fig. 6.2, one notices that the obtained hydrodynamic modes *coincide* with some of the modes of the original Grad's moment system. This observation reflects one of the basic tenets of the theory of invariant manifolds, already mentioned in this chapter, namely, that there is no loss of detail, i.e., of "information," in the transition from a microscopic to a projected, more macroscopic, description.

The analytical complexity of either the CE method or the invariance equations is overwhelming when we regard systems other than the linearized Grad's system or some linearized model of the Boltzmann equation, like those studied in Chap. 5. Approximate solutions are then the only feasible approach. We used, for instance, Newton's method, cf. Fig. 6.5, to solve iteratively Eqs. (6.19), taking as initial condition the Euler approximation (corresponding to a nondissipative hydrodynamics: $A_0 = \ldots = Z_0 = 0$), which leads, after the first iteration, to the same result, achievable alternatively through a technique of partial summation [6] of the CE expansion,

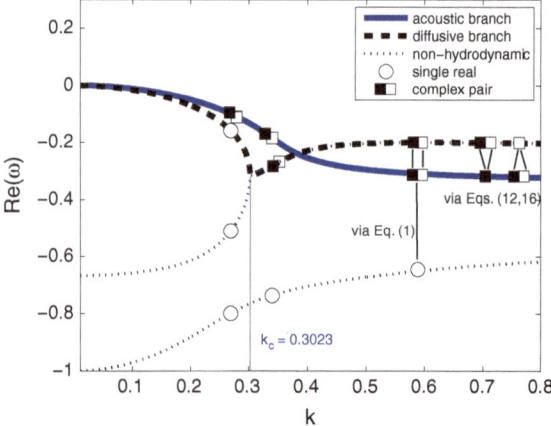

Fig. 6.2 Dispersion relation for the linearized 1D13M Grad's system (6.4). The unique solution of hydrodynamic modes obtained from (6.16) with (6.19) coincides with the real parts of the modes of the original Grad's system (the plot also shows when pairs of complex-conjugate roots appear). The solution of the original system (6.4) features five ω's, while the exact solution of (6.16) with (6.19) has three ω's for each k and is degenerate over the hydrodynamic branches at $k \geq k_c$

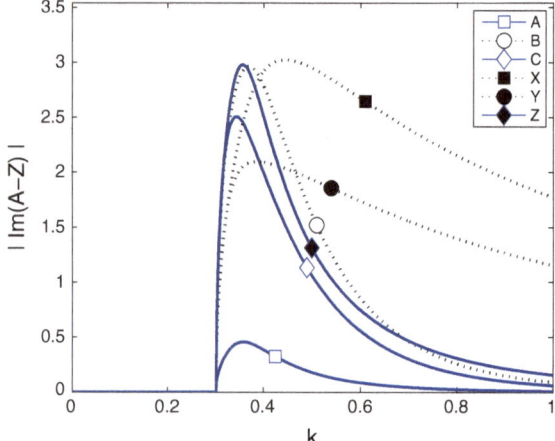

Fig. 6.3 Imaginary parts of coefficients $A-Z$ solving (6.19). Shown is the unique solution leading to hydrodynamic branches, cf. Fig. 6.2, which is symmetric about the real axis

essentially a sort of regularized Burnett approximation. It is seen in Fig. 6.5 that the Newton iterations converge rapidly to the exact hydrodynamics in the domain of its validity, $k \leq k_c$.

While we have evaluated the functions $A-Z$ numerically, two questions remain open: Is the stable exact hydrodynamics also hyperbolic? If so, how is it possible to retain hyperbolicity in the approximations? In the next section, we shall answer these questions.

Fig. 6.4 Real parts of complex-valued functions A–Z solving (6.19)

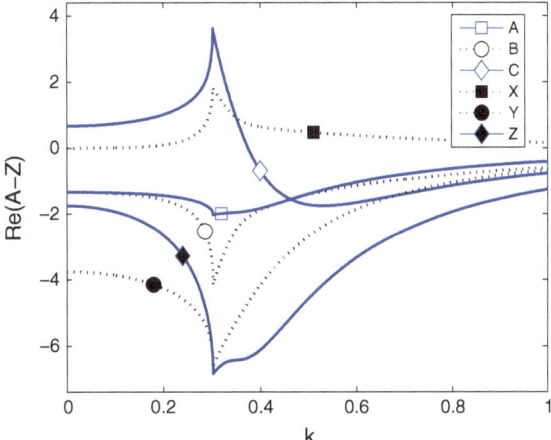

6.2.3 Hyperbolic Transformation for Exact Hydrodynamics

Equation (6.4) for the Fourier component vector $\mathbf{x}_k \equiv (\rho_k, u_k, T_k)$ reads $\partial_t \mathbf{x}_k = \mathbf{M} \mathbf{x}_k$ with $\mathbf{M}$ given in (6.15). By explicitly reintroducing the Knudsen number ε, i.e., by replacing k by $k\varepsilon$ in $\mathbf{M}$ and further distinguishing between the real and imaginary matrix elements in $\mathbf{M}$, we can write

$$\partial_t \mathbf{x}_k = [\text{Re}(\mathbf{M}) - i\,\text{Im}(\mathbf{M})]\mathbf{x}_k, \tag{6.20}$$

with

$$\text{Re}(\mathbf{M}) = \sum_{n=0}^{\infty}(-1)^n \mathbf{R}^{(n)}\varepsilon^{2n+1} = \varepsilon\mathbf{R}^{(0)} - \varepsilon^3\mathbf{R}^{(1)} + O(\varepsilon^5),$$

$$\text{Im}(\mathbf{M}) = \sum_{n=0}^{\infty}(-1)^n \mathbf{I}^{(n)}\varepsilon^{2n} = \mathbf{I}^{(0)} - \varepsilon^2\mathbf{I}^{(1)} + \varepsilon^4\mathbf{I}^{(2)} - O(\varepsilon^6),$$

and rearrange such that the Knudsen number expansion coefficients become visible. We find that the operators $\text{Re}(\mathbf{M})$ (real part) and $\text{Im}(\mathbf{M})$ (imaginary part) involve the following real-valued operators (for all $n \geq 0$, i.e., with the convention $a_{-1} = c_{-1} = z_{-1} \equiv 1$ and Kronecker symbol δ),

$$\mathbf{I}^{(n)} = k^{2n+1}\begin{pmatrix} 0 & \delta_{n,0} & 0 \\ b_{n-1} & 0 & c_{n-1} \\ 0 & \frac{2}{3}z_{n-1} & 0 \end{pmatrix}, \quad \mathbf{R}^{(n)} = k^{2n+2}\begin{pmatrix} 0 & 0 & 0 \\ 0 & a_n & 0 \\ \frac{2}{3}x_n & 0 & \frac{2}{3}y_n \end{pmatrix}. \tag{6.21}$$

The hydrodynamic equation (6.20) is hyperbolic and stable, provided that we can find a transformation of hydrodynamic fields such that

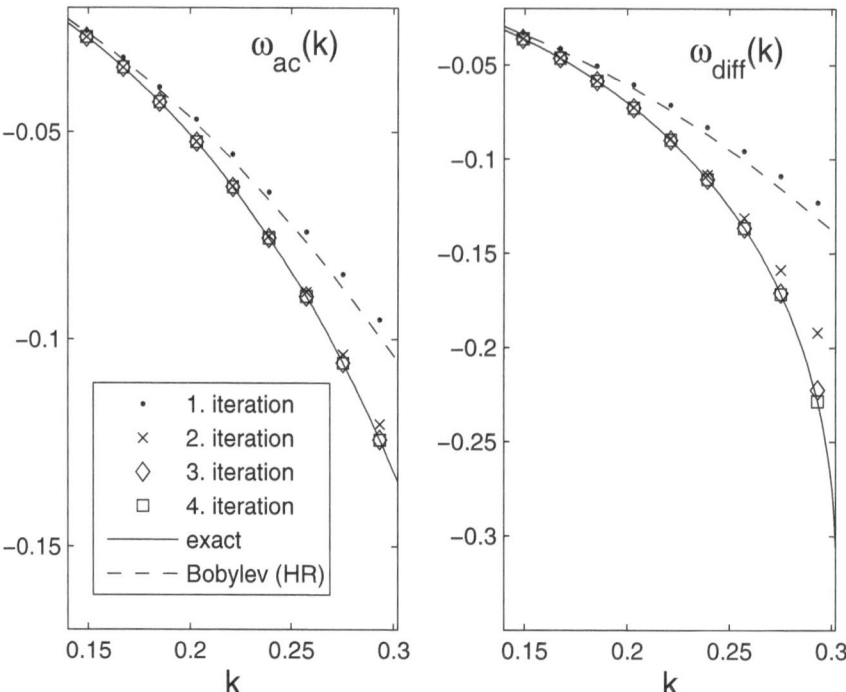

Fig. 6.5 Dispersion relations $\omega(k)$ for acoustic and diffusive modes obtained via Newton iteration. In the plots is also shown the approximation obtained through Bobylev's hyperbolic regularization. Newton iterations fail for $k \geq k_c = 0.3023$

(i) $\mathrm{Re}(\mathbf{M})$ and $\mathrm{Im}(\mathbf{M})$ are both real and symmetric,
(ii) $\mathrm{Re}(\mathbf{M})$ has negative semidefinite eigenvalues.

Therefore, we seek a transformation $\mathbf{x}'_k = \mathbf{T}\mathbf{x}_k$ that produces a symmetric matrix $\mathbf{M}' = \mathbf{T}\mathbf{M}\mathbf{T}^{-1}$, and we wish to see whether $\mathrm{Re}(\mathbf{M}') = \mathrm{Re}(\mathbf{T}\mathbf{M}\mathbf{T}^{-1})$ is negative semidefinite. We consider the equations of exact hydrodynamics, i.e., Eqs. (6.20), provided that functions $A, \ldots, Z$ (6.21) are solutions to the invariance equation (6.19). After a little algebra, which we do not recapitulate here, we obtain a particular transformation matrix $\mathbf{T}$ that solves the problem. It is a member of a whole class of effectively equivalent transformations, and can be written as

$$\mathbf{T} = \frac{1}{T_{uu}} \begin{pmatrix} T_{\rho\rho} & 0 & T_{\rho T} \\ 0 & T_{uu} & 0 \\ 0 & 0 & T_{TT} \end{pmatrix}, \tag{6.22}$$

with the nonvanishing components

$$T_{\rho\rho} = \frac{T_{uu}^2}{\sqrt{3X + 2Y[[Z]]}},$$

$$T_{uu} = \sqrt{X[[3B - 2Z[[C]] - 2C]] + 2Y[[B]][[Z]]},$$

$$T_{\rho T} = -\frac{3[[C]]X}{\sqrt{3X + 2Y[[Z]]}},$$

$$T_{TT} = \sqrt{3[[C]]\,(Y[[B]] - [[C]]X)}, \tag{6.23}$$

where we have introduced the following symbolic notation:

$$[[\bullet]] \equiv 1 - (k\varepsilon)^2 \; \bullet.$$

The transformation $\mathbf{T}$ (6.22) symmetrizes $\mathbf{M}$ and makes the system hyperbolic, as can be verified by a straightforward computation of $\mathbf{M}'$ from (6.22), (6.26). We further notice that $\mathbf{T}$ contains only even powers of $(k\varepsilon)$, because the same is true for the coefficients A–Z. Next we calculate the eigenvalues $\lambda_{1,2,3}$ of $\mathrm{Re}(\mathbf{M}')$—containing transport coefficients—to obtain a remarkably simple result:

$$\lambda_1 = 0, \quad \lambda_2 = k^2\varepsilon A, \quad \lambda_3 = \frac{2}{3}k^2\varepsilon Y. \tag{6.24}$$

From the analysis of the previous section, it follows that the nontrivial eigenvalues (6.24) are negative semidefinite for all $k\varepsilon$ (see Fig. 6.4, which displays the exact numerical solutions for A and Y). Hence, the equation describing the hyperbolic hydrodynamics attains the form

$$\partial_t \mathbf{x}_k' = \mathbf{M}'\mathbf{x_k'}, \tag{6.25}$$

with

$$\mathbf{M}' = \mathbf{TMT}^{-1} \tag{6.26}$$

for the vector $\mathbf{x}_k'$ of transformed hydrodynamic variables, and where $\mathbf{M}'$ is symmetric and has seminegative eigenvalues. To summarize,

$$\text{Hyperbolicity:} \quad \left(\mathbf{M}'\right)^T = \mathbf{M}', \tag{6.27}$$

$$\text{Dissipativity:} \begin{cases} \mathrm{Tr}[\mathrm{Re}(\mathbf{M}')] \le 0, \\ \det[\mathrm{Re}(\mathbf{M}')] \ge 0. \end{cases} \tag{6.28}$$

Equation (6.26) with (6.22) and (6.15) satisfying (6.28) is the main result of this section. The occurrence of negative eigenvalues in the exact solutions, together with the existence of a transformation $\mathbf{T}$ that makes the equations hyperbolic, proves that the exact hydrodynamics derived from (6.4), without approximations, is stable. Finally, we shall make use of the hyperbolicity of the exact hydrodynamics in order to establish approximate hydrodynamic equations that retain this property. In applications, one is interested in using truncated hydrodynamic equations by taking into account only lower orders of the Knudsen number ε. In this case, the functions A–Z are replaced by their lower-order approximations, and they can be generally written—as shown already in Eq. (6.11)—as polynomials truncated to an arbitrary order n. Their coefficients are usually derived through the CE recurrence equations, as outlined above. With the exact numerical solution in hand, we can also find, at any given order of approximation, the optimal interpolating functions A–Z solving Eq. (6.19), a method we wish to recommend and which has been worked out in [13]. The exact hydrodynamics, as described by Grad's system (6.4), terminates at k_c. In this regime, one can perform a Taylor expansion, up to any order n, on the elements of all three matrices $\mathbf{T}$, $\mathbf{M}$, and $\mathbf{T}^{-1}$. Thus, the approximations are done on the manifestly hyperbolic equation (6.25) in such a way as to retain hyperbolicity and stability at each order of approximation. It is worthwhile noticing that the eigenvalues, on approximating Eq. (6.25) to a polynomial order n, transform in a canonical manner:

$$\lambda_1^{(n)} = 0, \quad \lambda_2^{(n)} = k^2 \varepsilon \left(a_0 + \sum_{m=1}^{n} a_m (k\varepsilon)^m \right), \quad \lambda_3^{(n)} = \frac{2}{3} k^2 \varepsilon \left(y_0 + \sum_{m=1}^{n} y_m (k\varepsilon)^m \right),$$

and depending on the polynomial coefficients, and in particular depending on the sign of the highest-order coefficients a_n, y_n, the eigenvalues $\lambda_{2,3}$ diverge to $\pm\infty$ for $k\varepsilon \to \infty$, but stay negative for $k \leq k_c$ if one uses the set of coefficients reported in Table 1 of [13].

6.3 Exact Hydrodynamics from Three-Dimensional Linearized Grad's Equations

In this section, we extend the previous results to three-dimensional linearized Grad's equations [16]. The point of departure is the Fourier transform of the linearized three-dimensional Grad's 13-moment system

$$\partial_t \rho_k = -i\mathbf{k} \cdot \mathbf{u}_k,$$

$$\partial_t \mathbf{u}_k = -i\mathbf{k}\rho_k - i\mathbf{k}T_k - i\mathbf{k} \cdot \sigma_k,$$

$$\partial_t T_k = -\frac{2}{3} i\mathbf{k} \cdot (\mathbf{u}_k + \mathbf{q}_k),$$

$$\partial_t \sigma_k = -2i \overline{\mathbf{k}\mathbf{u}_k} - \frac{4}{5} i \overline{\mathbf{k}\mathbf{q}_k} - \sigma_k,$$

$$\partial_t \mathbf{q}_k = -\frac{5}{2} i\mathbf{k}T_k - i\mathbf{k} \cdot \sigma_k - \frac{2}{3} \mathbf{q}_k. \tag{6.29}$$

The goal is again to reduce the number of equations in (6.29) and to arrive at a closed set of equations for the hydrodynamic fields $[\rho_k, \mathbf{u}_k, T_k]$ only. To this end, we proceed as in Chap. 5 and decompose the vectors and tensors into parallel (longitudinal) and orthogonal (lateral) parts with respect to the wave vector, because the fields are rotationally symmetric around any chosen direction $\mathbf{k}$. We introduce, as shown in Fig. 5.1, a unit vector in the direction of the wave vector, $\mathbf{e}_\| = \mathbf{k}/k$, $k = |\mathbf{k}|$, and the corresponding decomposition $\mathbf{u}_k = u_k^\| \mathbf{e} + \mathbf{u}_k^\perp$, $\mathbf{q}_k = q_k^\| \mathbf{e}_\| + \mathbf{q}_k^\perp$, and $\sigma_k = \frac{3}{2}\sigma_k^\| \overline{\mathbf{e}_\|\mathbf{e}_\|} + 2\sigma_k^\perp$, where $\mathbf{e}_\| \cdot \tilde{\mathbf{u}}_k^\perp = 0$, $\mathbf{e}_\| \cdot \mathbf{q}_k^\perp = 0$, and $\mathbf{e}_\|\mathbf{e}_\| : \sigma_k^\perp = 0$. On inserting the above decomposition into (6.29) and using the identities $\overline{\mathbf{e}_\|\mathbf{e}_\|} \cdot \mathbf{e}_\| = (2/3)\mathbf{e}_\|$, $\overline{\mathbf{e}_\|\mathbf{e}_\|} : \overline{\mathbf{e}_\|\mathbf{e}} = \overline{\mathbf{e}_\|\mathbf{e}_\|} : \overline{\mathbf{e}_\|\mathbf{e}_\|} = 2/3$, we obtain the following two closed sets of equations for the longitudinal and lateral modes:

$$\partial_t \rho_k = -ik u_k^\|,$$
$$\partial_t u_k^\| = -ik\rho_k - ikT_k - ik\sigma_k^\|,$$
$$\partial_t T_k = -\frac{2}{3}ik(u_k^\| + q_k^\|),$$
$$\partial_t \sigma_k^\| = -\frac{4}{3}iku_k^\| - \frac{8}{15}ikq_k^\| - \sigma_k^\|,$$
$$\partial_t q_k^\| = -\frac{5}{2}ikT_k - ik\sigma_k^\| - \frac{2}{3}q_k^\|, \tag{6.30}$$

and

$$\partial_t \mathbf{u}_k^\perp = -ik\,\mathbf{e}_\| \cdot \sigma_k^\perp,$$
$$\partial_t \sigma_k^\perp = -ik\,\overline{\mathbf{e}_\|\mathbf{u}_k^\perp} - \frac{2}{5}ik\,\overline{\mathbf{e}_\|\mathbf{q}_k^\perp} - \sigma_k^\perp,$$
$$\partial_t \mathbf{q}_k^\perp = -ik\,\mathbf{e}_\| \cdot \sigma_k^\perp - \frac{2}{3}\mathbf{q}_k^\perp. \tag{6.31}$$

Equations (6.30) and (6.31) are a convenient starting point to derive closed equations for the hydrodynamic fields. To this end, the CE method amounts to eliminating the time derivatives of the stress tensor and of the heat flux in favor of spatial derivatives of the hydrodynamic fields of progressively higher order. As was shown in Sect. 6.1, we can express the stress tensor and the heat flux vector linearly in terms of the locally conserved fields by introducing six, as yet unknown, scalar functions $A(k), \ldots, Z(k)$ for the longitudinal part,

$$\sigma_k^\| = ikAu_k^\| - k^2 B\rho_k - k^2 CT_k, \quad q_k^\| = ikX\rho_k + ikYT_k - k^2 Zu_k^\|, \tag{6.32}$$

and two functions $D(k)$ and $U(k)$ for the transversal component,

$$\sigma_k^\perp = ikD\,\overline{\mathbf{e}_\|\mathbf{u}_k^\perp}, \quad \mathbf{q}_k^\perp = -k^2 U\mathbf{u}_k^\perp, \tag{6.33}$$

where the expressions for the longitudinal components share their form with the one-dimensional case, Eq. (6.4). Note that the functions introduced should be regarded as an exact summation of the CE expansion, which amounts to expanding these functions into powers of k^2 and deriving expansion coefficients from a recurrent (non-linear) system; cf. Eq. (6.9) and [6]. We do not dwell on this here, since we shall use a more direct way to evaluate functions $A, \ldots, Z, D, U$ in the sequel. Finally, using expressions (6.32) and (6.33) in (6.30), (6.31) and denoting by $\mathbf{x}_k = (\rho_k, u_k^{\parallel}, T_k, \mathbf{u}_k^{\perp})$ the vector of the hydrodynamic variables, the equations of hydrodynamics can be written in a compact form using a block-diagonal matrix $\mathbf{M}$:

$$\partial_t \mathbf{x}_k = \mathbf{M} \cdot \mathbf{x}_k, \qquad \mathbf{M} = \begin{pmatrix} \mathbf{M}^{\parallel} & 0 \\ 0 & \mathbf{M}^{\perp} \end{pmatrix}, \tag{6.34}$$

with

$$\mathbf{M}^{\parallel} = \begin{pmatrix} 0 & -ik & 0 \\ -ik(1-k^2 B) & k^2 A & -ik(1-k^2 C) \\ \frac{2}{3}k^2 X & -\frac{2}{3}ik(1-k^2 Z) & \frac{2}{3}k^2 Y \end{pmatrix}, \tag{6.35}$$

and

$$\mathbf{M}^{\perp} = k^2 D \begin{pmatrix} 1 & 0 \\ 0 & 1 \end{pmatrix}, \tag{6.36}$$

where the unit matrix is written in an (arbitrarily) fixed basis in the two-dimensional subspace of vectors $\mathbf{u}_k^{\perp}$. The matrix $\mathbf{M}^{\parallel}$, providing the evolution of the longitudinal modes, is exactly identical to the corresponding matrix $\mathbf{M}$ in (6.15) (relative to the one-dimensional case, in which lateral modes are absent). The twice-degenerate transversal (shear) mode is decoupled from the longitudinal modes. As a direct consequence, the invariance equations, to be discussed next, which will provide us with a set of nonlinear algebraic equations for the unknown functions A–Z, also divide into two subblocks, which can be solved separately.

Following the procedure outlined in Sect. 6.2.2, we introduce here the invariance equation for the 3D Grad's system (6.30) and (6.31). We find that the first set (six coupled quadratic equations for A, B, C and X, Y, Z) is identical to the set of Eq. (6.19). For the transversal modes, the invariance condition reads

$$\frac{\partial \boldsymbol{\sigma}_k^{\perp}}{\partial \mathbf{u}_k^{\perp}} \cdot \left(-ik\mathbf{e} \cdot \boldsymbol{\sigma}_k^{\perp} \right) = \partial_t \boldsymbol{\sigma}_k^{\perp}, \qquad \frac{\partial \mathbf{q}_k^{\perp}}{\partial \mathbf{u}_k^{\perp}} \cdot \left(-ik\mathbf{e} \cdot \boldsymbol{\sigma}_k^{\perp} \right) = \partial_t \mathbf{q}_k^{\perp}, \tag{6.37}$$

where the time derivatives on the left-hand sides of both equations in (6.37) are evaluated by the chain rule using $\partial_t \mathbf{u}_k^{\perp}$ given in (6.31). Substituting the functions (6.33) into (6.37), and requiring that the invariance condition be valid for every $\mathbf{u}_k^{\perp}$, we derive two coupled quadratic equations for the functions D and U, which can be cast into the following form:

Fig. 6.6 Real parts of coefficients A–Z solving the invariance equations (6.19) supplemented with (6.38)

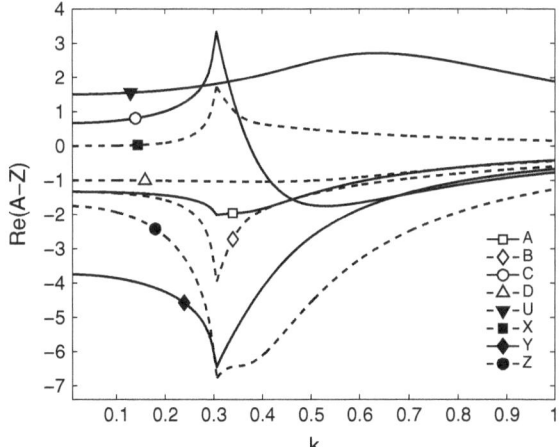

$$15k^4 D^3 + 25k^2 D^2 + (10 + 21k^2)D + 10 = 0, \quad U = -\frac{3D}{2 + 3k^2 D}. \tag{6.38}$$

The solution of the cubic equation (6.38) with the initial condition $D(0) = -1$ matches the Navier–Stokes asymptotics and was found analytically for all k. This solution is real-valued and is in the range $D(k) \in [-1.04, 0]$, whereas $U(k) \in [0, 2.72]$. The functions corresponding to the longitudinal part of the system were obtained numerically in Sect. 6.1. Because D and U are real-valued, we show in Fig. 6.6 the real parts for all coefficients, while their nonvanishing imaginary parts still coincide with those shown in Fig. 6.3.

The dispersion relations $\omega(k)$ for the five hydrodynamic modes are then calculated by inserting these coefficients into the roots of the characteristic equation $\det(\mathbf{M} - \omega\mathbf{I}) = 0$, where $\mathbf{I}$ is a 5×5 unit matrix.

Analogously, the dispersion relations for the remaining nonhydrodynamic modes follow from the eight (remaining) eigenvalues of (6.30), (6.31) with (6.32), (6.33). All 13 modes are presented in Fig. 6.7. The resulting hydrodynamic spectrum consists of five modes: the acoustic mode $\omega_{ac}(k)$, represented by two complex-conjugate roots; the real-valued thermal (diffusive) mode (both modes already occurring in the one-dimensional case); and a twice-degenerate real-valued shear mode (cf. Fig. 6.7). Just as in the one-dimensional case, a critical point in the hydrodynamic spectrum occurs at $k_c \approx 0.303$, where the thermal mode intersects a nonhydrodynamic branch of the original Grad's system. Hence the same conclusions hold here: for $k \geq k_c$, the CE method no longer recognizes the resulting diffusive branch as an extension of a hydrodynamic branch.

Figure 6.7 further shows the eight (all degenerate) nonhydrodynamic modes, which in contrast to the one-dimensional case (offering two nonhydrodynamic modes), also exhibit a critical k at $k'_c \approx 0.2175$. To summarize, the exact hydrodynamics as derived from an invariance condition (or equivalently, by the complete summation

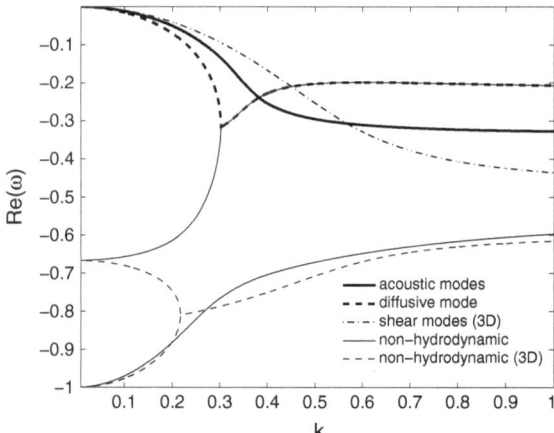

Fig. 6.7 Dispersion relations $\omega(k)$ for the linearized Grad's system using projected variables, Eqs. (6.30) and (6.31). The five hydrodynamic modes (diffusive, twice-degenerate shear, and two complex-conjugate acoustic modes), as well as the eight nonhydrodynamic modes, are presented as a function of k. While the acoustic mode is complex-valued for all k, the remaining modes become complex-valued beyond the two visible bifurcation points (at $k'_c \approx 0.2175$ and $k_c \approx 0.303$). For $k < k'_c$, the nonhydrodynamic (3D) modes are degenerate two and four times, respectively, corresponding to the two and four components of $\mathbf{q}_k^{\perp}$ and $\boldsymbol{\sigma}_k^{\perp}$

of the CE expansion) extends up to a finite critical value k_c, in full agreement with the one-dimensional case. No stability violation occurs, in contrast to the finite-order truncations thereof.

Next, we tackle the question concerning the hyperbolicity of exact hydrodynamics in the three-dimensional case.

Distinguishing between the real $(\mathrm{Re}(\mathbf{M}))$ and imaginary $(\mathrm{Im}(\mathbf{M}))$ parts of the matrix $\mathbf{M}$ (6.34), we can write the equation of the hydrodynamics conveniently as

$$\partial_t \mathbf{x}_k = [\mathrm{Re}(\mathbf{M}) - i\,\mathrm{Im}(\mathbf{M})] \cdot \mathbf{x}_k \tag{6.39}$$

$$\mathrm{Re}(\mathbf{M}) = \begin{pmatrix} \mathrm{Re}(\mathbf{M}^{\parallel}) & 0 \\ 0 & \mathbf{M}^{\perp} \end{pmatrix}, \quad -\mathrm{Im}(\mathbf{M}) = \begin{pmatrix} \mathrm{Im}(\mathbf{M}^{\parallel}) & 0 \\ 0 & 0 \end{pmatrix}. \tag{6.40}$$

The system (6.39) is hyperbolic and stable if, in analogy with the derivation outlined in Sect. 6.2.3, we can find a transformation of the hydrodynamic fields, $\mathbf{x}'_k = \mathbf{T} \cdot \mathbf{x}_k$, where $\mathbf{T}$ is a real-valued matrix such that for the transformed matrices $\mathbf{M}' = \mathbf{T}\mathbf{M}\mathbf{T}^{-1}$, we have that

(i) $\mathrm{Re}(\mathbf{M}')$ and $\mathrm{Im}(\mathbf{M}')$ are symmetric;
(ii) all eigenvalues of $\mathrm{Re}(\mathbf{M}')$ are nonpositive.

Due to the block-diagonal structure of (6.34) as well as to the fact that in Sect. 6.2.3 we solved the problem of finding a transformation with the desired properties for the one-dimensional case, the transformation exists also in the three-dimensional case, and it has the following form:

$$\mathbf{T} = \begin{pmatrix} \mathbf{T}^{\parallel} & 0 \\ 0 & \mathbf{T}^{\perp} \end{pmatrix}, \tag{6.41}$$

where $\mathbf{T}^{\parallel}$ is explicitly given by Eqs. (6.22)–(6.23) in terms of k, A–C, and X–Z, and

$$\mathbf{T}^{\perp} = \begin{pmatrix} 1 & 0 \\ 0 & 1 \end{pmatrix}. \tag{6.42}$$

Thus, the transformation $\mathbf{T}$ (6.41) symmetrizes $\mathbf{M}$ and renders the exact hydrodynamic equations manifestly hyperbolic. Furthermore, the transform $\mathbf{T}$ contains only even powers of k, because the same is true for the coefficients A–Z. The five eigenvalues λ_{1-5} of $\mathrm{Re}(\mathbf{M}')$ (or equivalently, of $\mathrm{Re}(\mathbf{M})$) are

$$\lambda_1 = 0, \quad \lambda_2 = k^2 A, \quad \lambda_3 = \frac{2}{3} k^2 Y, \quad \lambda_{4,5} = k^2 D. \tag{6.43}$$

From the analysis of the previous section, where we solved for coefficients A, D, and Y appearing in Eq. (6.43), cf. Fig. 6.6, it follows that all the eigenvalues λ_{1-5} are nonpositive for all k. Note that the matrix $\mathrm{Re}(\mathbf{M}')$ is diagonal with the diagonal elements (6.43).

Finally, the hyperbolic structure straightforwardly implies an H-theorem for the exact hydrodynamics. Note that due to the linearity of the system (6.4), the choice of a proper H-functional is not unique. We follow Bobylev [12] and consider an H-function—in terms of the transformed hydrodynamic fields—defined as

$$H = \frac{1}{2} \int \left[\rho'^2(\mathbf{r}, t) + u'^2(\mathbf{r}, t) + T'^2(\mathbf{r}, t) \right] d^3 r. \tag{6.44}$$

Here, the hydrodynamic fields $\mathbf{x}'(\mathbf{r}, t)$ are defined through the inverse Fourier transform of the fields $\mathbf{x}'_k$. Note that $\mathbf{x}'(\mathbf{r}, t)$ are real-valued because the real-valued transformation $\mathbf{T}$ is an even function of k, $\mathbf{T}(k) = \mathbf{T}(-k)$. Therefore,

$$H = \frac{1}{2} \int \left[\rho'_k \rho'_{-k} + \mathbf{u}'_k \cdot \mathbf{u}'_{-k} + T'_k T'_{-k} \right] d^3 k, \tag{6.45}$$

which, using shorthand notation, we can abbreviate as $H = \frac{1}{2} \langle \mathbf{x}'_k, \mathbf{x}'_{-k} \rangle$. Thus,

$$\begin{aligned}
\partial_t H &= \frac{1}{2} \left(\langle \mathbf{x}'_k, \partial_t \mathbf{x}'_{-k} \rangle + \langle \partial_t \mathbf{x}'_k, \mathbf{x}'_{-k} \rangle \right) \\
&= -\frac{1}{2} i \left(\langle \mathbf{x}'_k, \mathrm{Im}(\mathbf{M}'(-k)) \mathbf{x}'_{-k} \rangle + \langle \mathbf{x}'_{-k}, \mathrm{Im}(\mathbf{M}'(k)) \mathbf{x}'_k \rangle \right) \\
&\quad + \frac{1}{2} \left(\langle \mathbf{x}'_k, \mathrm{Re}(\mathbf{M}'(-k)) \mathbf{x}'_{-k} \rangle + \langle \mathbf{x}'_{-k}, \mathrm{Re}(\mathbf{M}'(k)) \mathbf{x}'_k \rangle \right).
\end{aligned} \tag{6.46}$$

Since $\mathrm{Im}(\mathbf{M}')$ is an odd function of k, terms containing $\mathrm{Im}(\mathbf{M}')$ cancel, and we have, owing to the fact that $\mathrm{Re}(\mathbf{M}')$ is an even function of k,

$$\partial_t H = \sum_{s=1}^{5} \int \lambda_s |x'_{s,k}|^2 d^3k \leq 0. \tag{6.47}$$

Thus, we have demonstrated, by a direct computation, the H-theorem for the exact hydrodynamics for $k < k_c$ (at $k = k_c$, the eigenvalues λ_2 and λ_3 become complex-valued) [16].

References

1. H. Grad, On the kinetic theory of rarefied gases, *Comm. Pure and Appl. Math.* **2**, 331 (1949).
2. D. Burnett, The distribution of velocities and mean motion in a slight nonuniform gas, *Proc. London Math. Soc.* **39**, 385 (1935).
3. A.V. Bobylev, *Sov. Phys. Dokl.* **27**, 29 (1982).
4. F. J. Uribe, R. M. Velasco and L. S. Garcìa-Colìn, Bobylev's instability, *Phys. Rev. E* **62**, 5835 (2000).
5. R. Balakrishnan, An approach to entropy consistency in second-order hydrodynamic equations, *J. Fluid Mech.* **503**, 201 (2004).
6. A.N. Gorban and I.V. Karlin, Invariant manifolds for physical and chemical kinetics, Lect. Notes Phys. **660** (Springer, Berlin, 2005).
7. I.V. Karlin, Exact summation of the Chapman–Enskog expansion from moment equations, *J. Phys. A: Math. Gen.* **33**, 8037 (2000).
8. I.V. Karlin and A.N. Gorban, Hydrodynamics from Grad's equations: What can we learn from exact solutions?, *Ann. Phys. (Leipzig)* **11**, 783 (2002).
9. A. N. Gorban and I. V. Karlin, Short-Wave limit of hydrodynamics: a soluble example, *Phys. Rev. Lett.* **77**, 282 (1996).
10. I. V. Karlin, G. Dukek and T. F. Nonnenmacher, Invariance principle for extension of hydrodynamics: nonlinear viscosity, *Phys. Rev. E* **55**, 1573 (1997).
11. I. V. Karlin, G. Dukek and T. F. Nonnenmacher, Gradient expansions in kinetic theory of phonons, *Phys. Rev. B* **55**, 6324 (1997).
12. A.V. Bobylev, Instabilities in the Chapman-Enskog expansion and hyperbolic Burnett equations, *J. Stat. Phys.* **124**, 371 (2006).
13. M. Colangeli, I.V. Karlin and M. Kröger, From hyperbolic regularization to exact hydrodynamics for linearized Grad's equations, *Phys. Rev. E* **75**, 051204 (2007).
14. A.N. Gorban and I.V. Karlin, Method of invariant manifolds and regularization of acoustic spectra, *Transp. Th. and Stat. Phys.* **23**, 559 (1994).
15. I. V. Karlin, A. N. Gorban, G. Dukek and T. F. Nonnenmacher, Dynamic correction to moment approximations, *Phys. Rev. E* **57**, 1668 (1998).
16. M. Colangeli, I.V. Karlin and M. Kröger, Hyperbolicity of exact hydrodynamics for three-dimensional linearized Grad's equations, *Phys. Rev. E* **76**, 022201 (2007).

Chapter 7
Conclusions

In this work, we employed the invariant manifold method to derive closed hydrodynamic equations from some kinetic models. The main novelty of our approach stems from the use of a nonperturbative technique that allows us to sum exactly the classical Chapman–Enskog expansion. The method postulates a separation between slow and fast moments, and allows us to extract the slow invariant manifold in the space of distribution functions.

A crucial aspect of our derivation that is not enjoyed by other techniques based on systematic coarse-graining procedures [1–3] is that the entropy production rate remains unaltered in the transition from the kinetic to the hydrodynamic level. The use of the thermodynamic projector, introduced in Chap. 3, makes it possible, in fact, to reduce the description without increasing the entropy production [4]. This entails, in particular, that one cannot derive irreversible macroscopic equations from reversible ones via the invariant manifold method. In this sense, the latter should be regarded as a reduction (or solution) technique, that is, a tool to solve irreversible equations obtained through coarse-graining methods.

The use of the thermodynamic projector leads to a refined description of the macroscopic evolution equations also at finite Knudsen numbers, and provides constitutive relations for the nonhydrodynamic fields that extend and complete the Navier–Stokes–Fourier model. While extending the description beyond the strictly hydrodynamic regime, one should not forget that our method relies on the notion of local thermodynamic equilibrium. The proposed technique is based, in fact, on the theory of normal solutions of the Boltzmann equation, introduced by Hilbert, which is a sensible approximation as long as the description is confined to length and time scales compatible with the existence of a local equilibrium. Interestingly, the computation of the power spectrum of density fluctuations, in Chap. 5, revealed that the obtained generalized hydrodynamic theory breaks down below a certain length scale of the order of the mean free path, in agreement with earlier results reported in the literature [5]. The equations of exact hydrodynamics, constructed by solving the invariance equation for the nonequilibrium distribution function, are hyperbolic and admit an

M. Colangeli, *From Kinetic Models to Hydrodynamics*, SpringerBriefs in Mathematics,
DOI: 10.1007/978-1-4614-6306-1_7, © Matteo Colangeli 2013

H-theorem. It is worth remarking that our solution of the invariance equation was considerably simplified by considering linear deviations from global equilibrium.

In our opinion, the basic assumptions adopted in this work, such as the separation of time and length scales, the existence of a local equilibrium, and the linearization around equilibrium, do not affect the relevance of our results. On the contrary, they rather witness that in simple enough models amenable to an analytic or numerical treatment, one of the basic problems of statistical mechanics [6, 7], i.e., the computation of the slow invariant manifold, can attain an exact solution.

References

1. A. N. Gorban, I. V. Karlin, P. Ilg, and H. C. Öttinger, Corrections and Enhancements of Quasi-equilibrium States, J. Non-Newtonian Fluid Mech. **96**, 203 (2001).
2. A. N. Gorban, I. V. Karlin, H. C. Öttinger, and L. L. Tatarinova, Ehrenfest's Argument Extended to a Formalism of Nonequilibrium Thermodynamics, *Phys. Rev. E* **63**, 066124 (2001).
3. A. N. Gorban and I. V. Karlin, Macroscopic Dynamics through Coarse-Graining: A Solvable Example, *Phys. Rev. E* **65**, 026116 (2002).
4. H. C. Öttinger, Betond Equilibrium Thermodynamics (Wiley, 2005).
5. B. J. Alder and W. E. Alley, Generalized Hydrodynamics, Phys. Today **37**, 56 (1984).
6. A. N. Gorban and I. V. Karlin, Thermodynamic Parameterization, *Physica A* **190**, 393 (1992).
7. A. N. Gorban, I. V. Karlin, and A. Yu. Zinovyev, Constructive Methods of Invariant Manifolds for Kinetic Problems, *Phys. Reports* **396**, 197 (2004).